THE MATERIAL GENE

The Material Gene

Gender, Race, and Heredity after the Human Genome Project

Kelly E. Happe

NEW YORK UNIVERSITY PRESS
New York and London

NEW YORK UNIVERSITY PRESS
New York and London
www.nyupress.org

References to Internet Websites (URLs) were accurate at the time of writing.
Neither the author nor New York University Press is responsible for URLs that
may have expired or changed since the manuscript was prepared.

LIBRARY OF CONGRESS CATALOGING-IN-PUBLICATION DATA
Happe, Kelly E.
The material gene : gender, race, and heredity after the human genome project / Kelly E.
Happe.
pages cm
Includes bibliographical references and index.
ISBN 978-0-8147-9067-0 (hardback)
ISBN 978-0-8147-9068-7 (paperback)
 1. Genomics—Social aspects. 2. Human genetics—Social aspects.
3. Genetic engineering—Moral and ethical aspects. I. Title.
QH438.7.H37 2013
572.8'6—dc23

                              2012048186

New York University Press books are printed on acid-free paper,
and their binding materials are chosen for strength and durability.
We strive to use environmentally responsible suppliers and materials
to the greatest extent possible in publishing our books.

Manufactured in the United States of America
10 9 8 7 6 5 4 3 2 1

Dedicated to the memory of Elizabeth Chase Lewis (1919–2003)

CONTENTS

ACKNOWLEDGMENTS

I count myself lucky that this book was mostly a labor of love and not just, well, *labor.* I have many people to thank for that. First, I'd like to thank Carol Stabile for many years of mentoring, support, and friendship. Her work on the intersection of feminism, technology, and political economy proved absolutely crucial to my thinking.

I continue to reap the benefits of the top-notch education I received at the University of Pittsburgh. Many thanks go to John Lyne, Gordon Mitchell, Robert Olby, Peter Machamer, John McGuire, and Danae Clark. In the ensuing years, I have benefited from colleagues and mentors who invested their time and energy in my intellectual and scholarly growth. I thank Karen Whedbee (especially for proving that the intellectual is not an endangered species in academia after all), Rob Brookey, Lois Self, and Beatrix Hoffman for much-appreciated feedback and support. At the University of Georgia, I thank Chris Cuomo, Patricia Richards, Susan Thomas, Bethany Moreton, Juanita Johnson-Bailey, Tom Lessl, and Roger Stahl for their inspiring scholarship and for their generosity in supporting my own work. Ed Panetta, my colleague in the Department of Communication Studies, was my intercollegiate debate coach many moons ago— I'm not at all surprised that I appreciate him as a colleague as much as I appreciated him as a mentor during those formative undergraduate days. Barb Biesecker has introduced me to both biopolitical scholarship and the people producing it—and this book has benefited as a result. Celeste Condit long ago inspired my interest in rhetoric and genetics; more recently, she has been a cherished mentor. I continue to enjoy our conversations about life, work, and, of course, genes. Finally, I thank Darrel Wanzer, Phaedra Pezzullo, Ron Greene, and Stuart Murray for their advice, engaging conversations, and influential scholarship.

Several students have served as my research assistants over the years. They include Ioana Cionea, Kristopher Cannon, Leland Spencer, Megan Fitzmaurice, and Sheng-yun Yang. Special thanks go to Brian Ray not only for excellent research assistance but also for hilarious impersonations of French theorists. (Don't forget about me when you're famous.)

I thank my students in the courses "Sex, Politics, Science, and Reproduction" and "Environmental Communication." Their curiosity, rigor, passion, humor, and activism made teaching a rewarding and edifying experience. Students in the graduate seminar on feminist rhetorical theory helped me refine my thinking on many of the concepts informing this book's analysis.

The research for and writing of the book was aided by summer research support at Northern Illinois University, which allowed me to develop my early thinking about race, gender, and genomics. At the University of Georgia, a UGARF grant provided the funds necessary for a one-course release in the spring semester of 2009 so that I could devote more time to writing. A fourth-year pretenure one-course release gave me additional time for finishing major revisions.

Clark Henderson, Carol Stabile, Jonathan Frye, Chris Cuomo, Barb Biesecker, and Roger Stahl read various sections of the manuscript and provided immeasurably helpful comments. Celeste Condit read the entire manuscript at a crucial stage, which resulted in a much stronger argument. My work also benefited greatly from several persons who generously gave me their time, including Larry Brody of the National Human Genome Research Institute, Barbara Balaban of the Long Island Breast Cancer Coalition, and many other research scientists and environmental breast cancer advocates.

Many, *many* thanks go to Lee Ann Pingel for editorial help during the final stages. Monica Casper, Lisa Moore, and the two anonymous reviewers provided thoughtful, supportive, and engaging feedback on the manuscript—this book is much improved as a result. The editors Ilene Kalish and Despina Gimbel provided not only helpful advice but also encouraging words at crucial stages. Responsibility for the final version, of course, rests with me alone.

My chosen family has been a source of support, encouragement, and good times. Heartfelt thanks go to Kelley Martin, Cricket Burwell, and Nick Rynearson for friendship and appreciating my raunchy sense of humor; to Susan Thomas for her generosity and for cheering me on; to Chris Cuomo and Karen Schlanger for friendship, intelligent conversation, and inventing the Prosecco play date; and especially to Patricia Richards for being such a great friend and colleague, demonstrating that it is all worth it in the end, and most importantly, for agreeing with me a lot.

I am especially lucky for my given family, which includes Jeff Happe, who helped with the transition back to Athens (and who, along with Molly Happe, introduced me to leftist politics at an early age); Pam Adriance, who made much-needed family time possible; and Brooke Ryan, who is the best auntie Dash could ever hope to have. My godmother, Cheryl Patrie, has been generous and kind and remains one of my biggest cheerleaders. The entire Barnard clan warmly welcomed me into their family with much holiday time respite and good cheer.

"It takes a village," they say, and I've had quite a nice one in Athens. I'd like to thank, from the bottom of my heart, the following people, who loved my son as if he were their own: John and Dana Butler, Gloria Huesser, Chelsea Woods, Madie Fischetti (of the infamous "Fischetti Five"), Elizabeth Hargrove, Patricia Richards, Oscar Chamosa, and, of course, my mother, the enormously talented Karen Lewis, whose generosity seemingly knows no bounds. She repeatedly made the long drive to Georgia to shower Dash with gifts, kisses, and cake, and she continues to model what it is to be a strong, intelligent, and principled woman.

I thank my partner, Clark Henderson, for love and companionship. His bravery in the face of some of life's most cruel challenges has been an inspiration. His thirst for knowledge, brilliant mind, and unshakable moral compass both challenge and ground me. Plus, he's fun to hang out with.

To close, I thank my sweet cherub of a son, Dashiell, whose attitude toward life is something along the lines of "Why walk, when you can skip?" You came into my life when I needed you most.

On January 11, 2008, I lost my good friend Chet Meeks to colon cancer. Chet was, without a doubt, one of the smartest, most talented scholars I've had the good fortune to know. I learned a great deal from him over the years. Chet was just thirty-two when his cancer was first diagnosed; he died two years later.

Chet did not lead an unhealthy lifestyle. He was not genetically predisposed to colon cancer. No one in his family had been diagnosed with colon cancer. The etiology of Chet's cancer was, and will remain, a mystery, although he often wondered about the years he spent living near the Hudson River, polluted with polychlorinated biphenyls (PCBs), when he was pursuing his PhD in sociology. *+ epi gene tics; mutations*

Determining ultimate causes is not, understandably, high on the list of one's concerns when diagnosed with cancer. Chet's goal was treating the cancer and curing it despite considerably bad odds. For the cancer patient, little emotional or intellectual space exists for theorizing one's disease in the midst of learning about the medical array of tests, treatments, and "cures."

Nevertheless, I know that Chet was, on some level, encouraged by the work I had been doing on genomics, epidemiology, and the politics of cancer, keeping conceptual and critical questions on the table, holding onto the necessity of theory in the midst of crisis—of real, in-your-face, material exigency. He was, after all, a social theorist. Although theory can be abstract, it is also an indispensable tool for making sense of the varied *theory* and often contradictory details of individual and collective embodied experience and everyday life. As Paula Treichler reminds us, "theory is not the creature disdained by . . . anti-intellectual traditions, including U.S. medicine, for whom *theory* is defined as that which is devoid of relevance for 'practice' and real-life experience. At the end of the day, *theory* is another word for *intelligence*, that is, for a thoughtful and engaged dialectic between the brain, the body, and the world that the brain and the body inhabit."[1]

Theorizing disease in this way—as engaged dialectic between brain, body, and world—is to ask how it is at once an experience of the corporeal body and also a historically variable concept that emerges from how we think about, discuss, and name disease, in both institutional and cultural contexts. This explains, for instance, how it is that the bodily experience of disease is largely determined by what its counterpart—normality or health—means. Moreover, the process of becoming a patient is invariably influenced by accepted understandings of race, gender, sexuality, and class status. Together, these bodily attributes and conceptual frameworks make some interventions possible, but not others, and these interventions in some cases can be matters of life and death. Theorizing disease and the body thus ultimately serves the pragmatic needs of medicine—including genetic medicine, which, we hope, will provide more humane cancer treatments that target somatic cancer cells—to alleviate suffering and tend to bodies existing at the interstices of health and illness, however contingent and contextually bound those concepts and states of being might be.

This book also explores what happens when we think of biomedical discourse as political. How, for example, does disease index the kinds of environments and social relations that we embody? Because bodily attributes like gender and race are laden with political and social significance, how is the process of becoming a patient simultaneously a biomedical phenomenon and a sociocultural and political one? And if the body is the site for the material convergence of social relations, how might it also be the site for social and cultural resistance, albeit without enacting the problematic logics of biomedicalization? Although epidemiology provides important and useful models for understanding the social dimensions of disease, it nevertheless is limited in what it can tell us about the world the body lives in. In the final analysis, epidemiology depends on the body as the site for action and not the political, economic, and social structures responsible for the collective experience of disease.

Heredity—a particular way of thinking about bodies, risk, and, by extension, the social and economic order—is fundamentally at odds with the notion that disease is the corporeal effect of embodied social relations. On the one hand, heredity, or the idea that one is born with various susceptibilities, hyperindividualizes and privatizes risk and disease. Genomic medicine predicts a future of what is called personalized medicine, wherein a patient's genome becomes the site of diagnosis and treatment of risk. On the other hand, hereditarian thinking imagines patients racially, which means that personalized medicine has the potential to become yet another privilege of white patients, for whom race does not enter into diagnosis and treatment practices.

Despite the paradox whereby genomics simultaneously reifies both individuals and populations, its immanent hereditarianism nevertheless effects a crucial displacement in both cases: disease is evidence of inherited defects, not embodied life. Genomics is, then, more than a disease paradigm—it is a political worldview that has been both a constant in US history and a particular way of performing ideological work during discrete moments in that history. Indeed, my interest in genomics began long ago with the question, what makes the genomic model of disease distinct from other models? The answer to that question emerged largely through a particular reading of eugenics. During a regrettable period of history in which a vicious racism and nativism intersected, however tangentially, with the newly discovered theory of heredity, eugenics justified—in the minds of reactionaries and progressives alike—the figurative and literal criminalization of blacks, women, immigrants, and poor people. Why, I ask in this book, has heredity remained uniquely suited to the task of constructing biopolitical discourse during crucial and sometimes painful phases of industrial capitalism and emerging social movements and discourses of resistance?

To understand why it matters that Chet was stricken with an aggressive cancer at such a young age, we must think beyond his individual body and the treatments it required. We must think of his experience through the framework of collectivity, of his body located in space and time with others, living in environments and social relations not of their own making but resulting from deeply politicized and self-interested corporate, governmental, and institutional practices. To answer the questions I pose requires the humanist's eye, located a safe and critical distance from the disciplinary norms of genomics, medicine, and public health.

In the spirit of scholarly inquiry, and in memory of Chet's unrivaled intellect and sense of humor, I present the following study.

1

Ideology and the New Rhetoric of Genomics

In a 1999 article in the journal *Plastic and Reconstructive Surgery*, a surgical team describes the case of "A.H.," a patient who undergoes an eleven-hour operation to remove her breasts, ovaries, and uterus. The surgery also included the reconstruction of breasts using skin and tissue from various parts of her body. In all, the surgery involved three separate surgical teams and was divided into four stages. After a four-day hospital stay, A.H. was released; she underwent additional procedures on her breasts over the next seven months.

Although narrated in the matter-of-fact clinical language of the case report, the surgery was, in fact, radical, even gruesome, and the article included photographs of her body from the neck down with the customary markings of the plastic surgeon, noting the location of various incisions and cuts necessary to refashion her physique. Such surgery requires, the report informs us, "significant recovery time."[1] The authors of the report predict that the procedure will extend the life of A.H. by four to six years "or more," even though the decision model they employ assumes that there is a complete mastectomy and no reconstruction (which in fact A.H. had), and that the patient tolerates, and complies with treatment during many years of potentially dangerous hormone replacement therapy.

What sort of disease would necessitate such interventions? As it turns out, no disease at all, at least not in the conventional sense of the term. A.H. did not have cancer, nor did she have any symptoms of diseases related to the reproductive organs. The surgery was attributed, rather, to a positive test for a mutation of one of the so-called "BRCA" genes—in this case, a mutation of BRCA2.[2] Inheriting a mutation of one of these genes increases a woman's risk for breast and ovarian cancer. A.H.'s surgery was performed for the sole purpose of reducing her risk for both. Many

>> 1

women who have publicly shared their experiences with the test and subsequent surgery say that the BRCA test empowered them to make life-saving decisions. Prophylactic surgery does decrease risk for some women,
and those who choose this surgery do so with the knowledge that cancer
treatments can inflict far more violence on a woman's body (knowledge
they often gleaned as they witnessed the struggles of loved ones). And as
far as ovarian cancer is concerned, most women do not have the privilege
of considering treatment with any likelihood of success—over 60 percent of women diagnosed with the disease do not survive it. Mindful of
the concerns of bioethicists, advocates of BRCA screening claim that expanded genetic counseling and restrictions on direct-to-consumer marketing can help prevent the unnecessary administration of BRCA tests. In
the long term, they say, the genomics revolution in women's cancer care
will lead to better detection and prevention and to less invasive treatment,
both for the small percentage of women with BRCA mutations and those
without them.

   This perspective on testing and surgery, although important to the development of best practices in law and medicine, is ultimately limited by
the pragmatism on which it depends for coherence. Following in the traditions of rhetoric, feminism, and cultural studies, the present book asks
instead, how do we theorize and historicize medical discourse and practice amid the material reality of disease that otherwise compels us to act?
How do we think of heredity and disease not as recalcitrant material realities discovered by researchers and physicians, but as contingent manifestations of the lived experience of social worlds? Why does the integration
of genomics into medicine and public health in particular require us to
think of disease in this way? And what methodological tools are necessary
and sufficient for the task?

   As these questions suggest, I take the story of A.H. to be more than
one woman's struggle with cancer prevention; it is, rather, an artifact of a
larger social discourse about the body and the world that body inhabits.
Epistemologically and rhetorically, genomics effects the dematerialization
of the body and its embeddedness in historically specific environments
when biological matter is translated into the language of gene sequences
and risk assessment. The body becomes just so much information. Yet the
body rematerializes when genomics must make fathomable and palpable
the body at risk, both to fashion medical subjects and to engage in the
making of procedural rules and norms necessary to formalize and rou-
tinize particular sets of interventions. Because bodies are always already

culturally inscribed—Katherine Hayles calls this the outcome of ideological transcription practices[4]—genomics is dependent on normative conceptions of the gendered, raced, and bounded body. As Catherine Waldby reminds us, "Biomedical knowledge cannot . . . be quarantined from general ideas operative in the culture, even when it understands its concepts to be carefully and directly deduced from the factual evidence of the body. Despite, or perhaps because of, biomedicine's assertion of its own innocence of historical and political meaning, it constantly absorbs, translates and recirculates 'non-scientific' ideas—ideas about sexuality, about social order, about culture—in its technical discourses."[5]

Not only must genomics rely on cultural discourses about the body to translate genetic information into body practices, but in doing so it will in turn participate in their construction. It thus contributes to our shared meanings of race, gender, and embodied life. Indeed, the very conditions of intelligibility on which the story of A.H. depends in both popular and scholarly discourse rest on hidden, and heretofore underexamined, assumptions about gender, race, and political economy. Family history aside, A.H., through the visualization and objectification of her body, is removed from any context that might allow for a consideration of why the female medical subject figures so prominently in the genomics revolution; how gender and race determine whether the removal of her organs is the rational way to prevent cancer; whether her genomic profile is simultaneously a racialized one; and whether the risk she fears is appropriately construed as an inborn trait.

We must interrogate these otherwise mystified embodied materialities to understand not only how the normative body mediates the translation of inherited risk into actual biomedical practices, but also how biomedical theories and the social and economic order inform each other. The chapters that follow thus recontextualize the body of A.H. and theorize its materiality within a cultural, economic, and political context. In so doing, the book explores genomics' implicit investment in the very norms and values on which its translation into medicine and public health depend. A central argument of the book is that explicit ideological appeals are no longer required for genomics, and biomedicine more broadly, to shore up the values of the free market, the racism of biological types, and the antifeminism of the conflation of gender and sex. Rather, the means by which genomics performs this ideological work (and, indeed, how it serves its own institutional interests) has along with shifts in gender and racial politics, social movements, and political economy. I attempt to explain this ideological work by mapping—and problematizing—a logic of progression emerging from developments both internal and external to genomics.

Genomics, the Body, and the Economic and Social Order

In this book, I examine a variety of research projects on women and cancer in order to reconsider genomics' relationship to its eugenics past. By doing so, I challenge the received view that the science and politics of eugenics can be neatly circumscribed within a particular historical era and that any residual effects of its legacy can be attributed to a few individual scientists who continue to search for the biological inferiority of women, blacks, and other minority groups. I argue, instead, that a hereditarian ideology and concomitant regressive politics has been a historical constant, informing various incarnations of genetics over the past hundred years.

My analysis begins in the early twentieth century, framing eugenics as an early example of genomics' explanatory capacity to attribute the causes of material and social inequalities to the pathologies of particular bodies. During the progressive era, the scientistic worldview of genetics cast the biological sciences as salvation from the extreme poverty, civil unrest, and overall exploitation associated with industrialization; the material consequences of the Industrial Revolution, so the narrative went, were really the products of unfit breeding. Of particular significance was the fact that middle-class consumers of eugenics discourse could—like their middle-class professional counterparts in the sciences—espouse a worldview that identified them neither with the upper class nor with workers, the poor, or immigrants. Instead, genetic engineering, whether through positive or negative eugenics (propagating or violently repressing particular traits), could produce better behaved leaders and followers alike. Although the critique of unfettered capitalism was a defining feature of eugenics discourse, the latter nevertheless recuperated the fundamental values of the free market by displacing the struggle for redistribution of resources with the technological interventions of science. The eugenics period thus provides an early example of how genetics advanced its methodologies (for example, by taking seriously for the first time theories of Mendelian inheritance) while simultaneously taking part in the formation of a depoliticized, dehistoricized view of social change.

It is this feature of eugenics discourse—not the specific claims it made about the science of heredity or the pathological behaviors it could explain—that necessitates its figuring in any critical account of genetics today. As I will show throughout the book, genetics never abandoned its investment in the normalization of bodies and the larger social and economic order, even as it set out to reinvent itself after the demise of the eugenics movement. However, since genomics is no longer articulated to explicit political projects like eugenics and its close cousin social Darwinism, we must consider how hereditarian ideology is located in what Sandra Harding, following Fredric Jameson, calls science's "political unconscious."[6] For Harding, science's cognitive core—the collection of theories, models, tools, and norms that guide its practices—is never merely that. Rather, embedded within those practices is a politics, an investment in a particular arrangement of social relations outside the confines of the research setting. Responding to what she says is the myth that "inegalitarian political projects remain external to fundamentally value-free scientific assumptions, methods, and claims," she argues that science is always already ideological, even when it passes the highest standards of methodological excellence and explicitly distances itself from culture, politics, and political economy.[7] A closely related concept is Sheila Jasanoff's notion of "co-production," a novel take on knowledge making in scientific research that "calls attention to the social dimensions of cognitive commitments and understandings, while at the same time underscoring the epistemic and material correlates of social formations."[8] Thus science is also a form of social knowledge,[9] but with a set of epistemological practices and discourses unique to it.

The concepts "political unconscious" and "co-production" together capture the complex relationship that institutions have with each other and with the social and economic order.[10] These relations are not always direct or easily mapped. Instead, they manifest themselves as a form of ontological or ideological complicity—separate fields of discourse that are semi-autonomous (and so are characterized by epistemological practices unique to them), but nevertheless permeable to and thereby shaped by historically resilient dominant interests and their cultural logics.[11]

Throughout the book I show how, through an institutional discourse of health and healing, the political unconscious of genomics—its hereditarianism—has become more difficult to locate and map but nevertheless remains a powerful method for naturalizing social and economic relations. Human genetics' insinuation into the provinces of medicine and public health has been one of the most effective means by which it claims to be free of such interests. Having retreated to the world of the laboratory after the demise of the eugenics movement, genetics was once again revital-ized with the publication of new research demonstrating the heritability of disease, now more narrowly defined to exclude the behavioral traits fa-vored by its eugenics predecessors. More recently, the Human Genome Project (HGP) has promised to revolutionize medical and public health practice by making heredity central to understanding and modifying dis-ease etiology.

Nevertheless, even though genomics is concerned principally with dis-ease, and not with the behaviors or political leanings of its subjects, it is no less invested in particular economic and social arrangements. My argument rests on the notion that disease exceeds, both symbolically and materially, mere physiological changes that distinguish the normal from the pathological state of the body—the absence or presence of a breast cancer tumor, for instance. Rather, disease may be said to reveal how op-pression gets "under the skin."[12] Class, for instance, is the "best predic-tor of life expectancy, of old-age disability, or [of] the frequency of heart attacks. As a predictor of coronary disease, it is better to measure class position than to measure cholesterol."[13] The lived experience of race is very much connected to hypertension and particular types of cancers. Gender explains how experiences in the natural and social world influ-ence the kinds of diseases women are more likely than men to struggle with; furthermore, it overdetermines the kinds of treatments that women will be offered.

This model of disease as the materialization of embodied life is dis-placed—conceptually and in terms of research dollars—by genomics' explanation of disease patterns as the product of genes.[14] The genes with which genomics are concerned are, theoretically, alleles of populations, not sexes or races; but because the object of genomics in medicine is the body, heredity inevitably becomes bound up with these very catego-ries. The specific displacement that obtains in genomics discourse is the privileging of a diachronic understanding of race and gender insofar as it assumes these bodily attributes to be traits that are passed on from one generation to the next. [15] Even when gene function is qualified to include the role of environmental factors, hereditarianism typically compels a diachronic analysis of the body and disease: genes are inherited at birth, and the evolutionary history of this transmission is what substantiates bodily narratives. And to a large extent, the stability of genetic substance (there are few genes, relatively speaking, in human DNA, and the du-rability of haplotypes is what has made genetic mapping even possible) overdetermines meanings of the gene, even in those models that purport to give due consideration to the agency of environmental contexts.

The synchronic view, in contrast, holds that gender and race are descriptors of lived and embodied experience, and as such are inextricably bound with political, social, and economic structures.[16] A synchronic view of race, for example, would require an examination of social isolation, geography, and what critical race scholars call "microaggression"—the relentless exposure to small acts of racial prejudice on an everyday basis.[17] And it would treat these variables as historically contingent, materializing in ways particular to a given place and time. What does it mean, for example, to embody racial identity in a specific geographic location, during a particular stage of capitalism, and within historically specific practices of institutional racism? And with these questions in mind, what does it mean to embody sex?

These different models of the body and disease matter not only because of the kinds of biomedical practices they call forth (what science studies scholars have called the inextricable link between representing and intervening)[18] but also because describing bodily attributes in synchronic terms makes possible a particular kind of agency and accountability—namely, a call for changes in economic and social policy as a way to effectively deal with health inequities. Genomics' privileging of the diachronic over the synchronic thus reveals a logical homology between its practices and the political discourses that rely on such a displacement. As Jasanoff says of "co-production," the "ways in which we know and represent the world (both nature and society) are inseparable from the ways in which we choose to live in it."[19]

Genomics is by no means alone in privileging a diachronic reading of the body and disease—I will show that, in many ways, it has acted as the successor to epidemiological discourse that has trafficked in the idea that the body can be broken down into proximate, isolatable, and static components. Indeed, biopolitics scholars have quite persuasively questioned the very notion of health. Genomics has not, however, enabled the types of progressive discourses that epidemiology has.[20] Heredity, I will suggest, is overdetermined by its eugenics legacy—a cultural logic that translates the pathologies of economic and social relations into pathologies of bodies—inherited at birth and immune to change.[21] The chapters of the book together examine how this logic is introduced and what is at stake by asking: How do genomics and its implicit hereditarianism enter into long-standing debates about women, reproduction, and the ideology of motherhood, specifically attitudes about women's access to education and economic independence? How does the diseased ovary become both an immutable trait and a signifier of dominant ideals of womanhood and femininity? How are those same ideals transformed into pathological racial attributes, whereby black mothers are blamed for material conditions not of their own making? How is genomics' explanation of health disparities both an expansion of its institutional interests and an implicit investment in an economic order that is otherwise threatened by the growing fields of social epidemiology and critical health studies, both of which implicate institutional racism, not race, as the source of those very disparities? And how is the geneticized subject of public health also a neoliberal subject insofar as it valorizes personalized, privatized health interventions at the expense of progressive environmental policy reform?

In considering how the genomic model of the body is logically homologous to, and complicit with, the interests of the status quo investment (both literally and figuratively speaking) in hierarchical social relations and the concentration and privatization of wealth, my analysis is similar to that of feminist body studies showing logical and discursive homologies between models of biological systems and those of communication systems, global circulation of capital, and flexible accumulation. Emily Martin, for example, has described the "Fordist body"—the rhetorical product of medical discourse during this particular phase of industrial production—and, more recently, has examined how the idea of "flexibility" informs immunology, production practices, and the ideal body.[22] Flexibility, in short, has both ontological as well as epistemological dimensions.

Similarly, Donna Haraway has described evolving models of the immune system, models that bear a striking resemblance to concomitant systems of production and communication unique to late capitalism, such as flexible accumulation and just-in-time inventory.[23] Melinda Cooper, updating Foucault's work on the homologies linking political economy and the biological sciences, has shown how various projects of the biosciences (for example, stem cell research) are neoliberal insofar as they materialize the ideals of surplus, speculation, and globalization.[24]

This work has described representations of biological systems and their constituent parts in biomedical discourse. But what about actual body practices? What is the relationship between discourse and the possibilities for the lived body? Biological and medical discourses reveal the fact that scientists are embodied persons whose language practices reflect a temporal and spatial positionality, but *how* does biology act as a discursive space wherein the historical, material, and contingent conditions of capitalism are visualized, and eventually internalized, as the natural order of things?"[25] Donald Lowe, in *The Body in Late-Capitalist USA*, says that bodily practices reflect the changing needs of the market and are ways in which its values become part of common-sense, everyday, embodied life. For example, deskilling changes the material ways in which the body produces surplus value, by radically constricting and accelerating one's movements; sexuality and gender become commodifiable "lifestyles."[26] By drawing from both Marxist and discourse methodologies, Lowe compellingly shows the advantages of a synchronic analysis of the body: bodily practices are enabled by specific configurations of discourse and of production and consumption practices, the latter being unique to the stage of capitalism in which the analysis proceeds.

But the circulation and imposition of these values will not affect all bodies in the same way. As Waldby observed in her study of AIDS discourse, body practices are a mechanism for the enforcement of social norms. Thus, we must attend to the specific ways in which systems of gender and racial oppression are called on to discipline bodies.[27] In this way, racism and sexism are understood as social phenomena distinct from economic structures and relations (and so not reducible to them), but nevertheless integral to their successful operation.[28] Regarding race, Lowe writes that its context "cannot be analyzed in terms of an orthodox Marxist structural order, but as a hegemonic terrain of exchangist practices, with the recombination of structural, discursive, systematic, and semiotic components. And the very structural, discursive, systematic, and semiotic components are premised on the pervasiveness of racism. In other words, race operates within a hegemonic terrain, and that terrain takes advantage of existing racism. In reality, the two are intertwined." Race is "neither a monolith, nor an isolatable variable," he continues. "Therefore, let us abandon the search for an isolatable racism which, in effect, frees all other variables from any responsibility," for economic interests are often secured by exploiting "existing racial stratification and racism."[29]

Lowe's analysis, in providing a synchronic account of bodily practices during the mid-1990s, both shows us how the body is the site for the production of surplus value and, in so doing, also shows us how body scholarship can help explain particular stages of capitalism. In this book, I too want to look at the lived body in late capitalism (or, more precisely, neoliberalism), but not as a way to understand the body as a site for capital accumulation per se (that work has been done already).[30] Rather, taking biomedical discourse as a crucial space for analysis, I juxtapose the genomic model of the body—what I have called, following Jacqueline Stevens, a diachronic perspective—with models of the lived body in social and economic relations. It is thus an account of how the relations between biomedicine, political economy, and hierarchical social structures are mystified through the practices of genomics—namely, by producing discourses of medicine and public health that enable its subjects to embody the very norms and values on which the economic order depends.

More specifically, the book adds to the extant literature on body studies and biosociality, a feminist materialist analysis that presumes the "Fordist" body and "late capitalist" body are raced and gendered. Dominant racial and gender norms influence the types of biosocial subjectivities, relations, and practices that medicine and public health make possible. The chapters of the book thus attend not only to the links between hereditarianism and the needs of capital, as I have summarized above, but also to how other systems of oppression inform, and make possible, the particular knowledge claims of genomics. In ovarian cancer genomics, for example, the intersection of medicine with patriarchy reduces women to reproductive function and desire, a reduction that historically has served to discipline women and force them to adhere to limited, exploitative gender roles. The population turn in breast cancer genomics, emerging within a history of racism and racialism in science and medicine, is also a story of how genomics, constituting African American women as racialized subjects of genetic testing, helps serve the interests of the state in defunding antiracism programs. And class alone cannot explain the phenomena of environmental breast cancer and pollution in communities of color. Nevertheless, race and gender mediate the emergence of the neoliberal subject at the center of the incorporation of hereditarianism into public and environmental health.

Rhetoric, Ideology, and Progress

Not all scholars of biomedicine and genomics agree that a political unconscious underlies the theories and methods of science. Specifically, the convergence of assorted interests around biomedicine has led some science studies scholars to conclude that heredity cannot and does not determine the outcome of genomics' basic claims. Nikolas Rose argues, for instance, that it is impossible to locate an "intrinsic politics" within genomics. Regarding the specific matter of race, he says:

As with [sickle cell anemia], so more generally in the relations between race and genetics: these links have no given or intrinsic politics; they take very different forms as they are entwined with distinct
styles of thought about health, illness, and the body at different times
and places. And, as with [sickle cell], the relation between these terms
today, in the United States at least, is intrinsically linked to the delinea-
tion and administration of biosocial communities, formed around beliefs in a shared disease heritage, demanding resources for the biomedical research that might reveal the genomic bases of these diseases, and
mobilized by the hope of a cure. Perhaps, then, we might understand
the contemporary allure of race in biomedicine in terms of the hopes,
demands, and expectations of such communities of identity, as both
subjects and targets of a new configuration of power around illness and
its treatment.[31]

There is little doubt that the allure of race is indeed a product of the
collective demand of various communities of persons that biomedicine
attend to, among other things, disparities in the incidence and death rates
of various diseases. The problem with Rose's overall argument, however,
is that he mistakes the desire for a progressive genomics with the possibilities for one. Put another way, inclusivity does not necessarily entail
a corresponding shift in political or ideological alignments. The constel-
lation of varied agendas is not, I would argue, evidence that the science
itself, and its technological interventions, are value-free, undetermined
by the conditions of their emergence.[32] Certainly, genomics' attention to
race may appear to satisfy many interests simultaneously, especially those
of African American members of the medical and research communities.
Yet this convergence can coexist with an otherwise undisturbed system of
racial oppression.[33]

And this system of oppression, as well as others, becomes more difficult to discern. Carol Stabile, in her work on the intersection of feminism,
technology, and political economy, has argued that dominant interests
can accommodate and withstand apparent challenges to their hegemony.[34]
In fact, they often must adapt to new material and discursive conditions,
and this adaptation does not always require recourse to older ideological
scripts. Ideological contradictions, she writes, are not necessarily resolved
in favor of "tradition," but resolve or repress "ideological contradictions
through a logic of progression."[35] Perhaps nowhere is this more apparent
than in the field of medicine, which earlier in its history very explicitly

made connections between its theories of the body and popular or lay notions of the inferiority of women, blacks, and the working class. For the most part, these sorts of explicit ideological appeals are no longer available for adoption and circulation in biomedical discourse.

With this in mind, genomics' attention to racial disparities in health may very well be the result not only of what cannot be said about race but also of what *must* be said about race in the contemporary moment. On one hand, a consensus has long existed in the biological and social sciences that race does not exist, no matter what criteria (genes, blood types, and so forth) are employed. On the other hand, growing attention to health disparities linked to one's racial identity has ensured that race, understood as a social phenomenon, remains an important object of scientific inquiry. Genomics' participation in this field is not so much a sign that racialism and racism have disappeared as it is an indication that new constraints on their articulation must be negotiated.[36]

Beyond the specific case of race, there are a number of developments internal and external to genomics that mystify its ideological commitments. For example, various theories of biological development have cast doubt on the centrality of DNA, prompting nothing short of a nonessentialist, postmodern reading of body processes as dialectical, contingent, and fluid.[37] We now know, for example, that the human genome contains far fewer genes than previously thought and that their actions and products are varied and environmentally conditioned. Several genes can code for the same protein, and just one gene can code for several different kinds of proteins. This is because protein synthesis is shaped by cellular, as well as extracellular, environments. Even if one concedes that DNA functions as a kind of "code," cellular mechanisms "interpret" DNA, thus qualifying its agency considerably. The genome is acted on as much as it acts; these theories have conceptually decentered DNA despite popular representations to the contrary.[38] (More radical still is the epigenetics' claim that somatic changes can be inherited.)

These developments, internal to genomics, make it possible for bodies to be no longer defined by, or reduced to, DNA. Instead, genomics research increasingly employs theories of complex patterns of allele frequencies and gene-environment interactions, as well as the revived theory—once attributed exclusively to Soviet genetics—that our material lives, including their historicity and contingency, shape what it means to be human.[39] Donna Haraway has observed that "ideologies of human diversity have to be developed in terms of frequencies of parameters and fields of power-charged differences, not essences and natural origins or homes. Race and sex, like individuals, are artifacts sustained or undermined by the discursive nexus of knowledge and power. Any objects or persons can be reasonably thought of in terms of disassembly and reassembly; no 'natural' architectures constrain system design."[40]

Likewise, developments external to genomics—such as the social changes brought about by civil rights, feminist, and environmental movements—have compelled significant discursive shifts. Gynecological oncologists are careful to qualify treatment recommendations that may conflict with the reproductive autonomy of women, a development that reflects the influence of the women's health movement. Moreover, opposition to ovary removal is not likely to be based on the notion that it "desexes" or "defeminizes" women, a reflection of the success of feminism in disarticulating sex from gender identity. Manipulating the body at risk can be reasonably interpreted as a way to distance oneself from a naturalized, inviolable understanding of female sexual identity. The postgenomics body becomes a cyborg of sorts: one with prosthetic breasts and chemical hormone supplements; and one that embodies flexible, contingent notions of environmental risk.

The emergence of influential women's and African American health movements, moreover, has meant that all fields of biomedicine, including genomics, feel a responsibility to attend to the health disparities that continue to befuddle epidemiologists. In many ways, developments in biomedicine—and genomics, more specifically—are seen as welcome changes to medicine's historical unwillingness to address gender and racial disparities, a failure that has contributed to social inequalities between women and men and between blacks and whites. And finally, a growing awareness of the links between environmental pollution and disease has given new importance to the gene-environmental model of disease etiology. Environmental breast cancer activists, for example, have repeatedly called on researchers to investigate the complex origins of tumors, including the role of genes in increasing one's susceptibility to toxic chemicals. These discourses, still unfolding, have produced bodies seemingly devoid of genetic or cultural essences but that nevertheless require continued study and care.

This logic of progression pervades much of the discourse of genomics: genetic testing can save women's lives; gene-environment interaction is a model for improving public health, not just the health of those with rare genetic disorders; and the genome provides material evidence of the universality of DNA. Yet as I argued earlier in the chapter, biomedicine, as an institution, both reflects and actively participates in the production of a particular gendered, racialized, economic order. One nodal point of this intersection of interests is the class position of scientists themselves. Historically, geneticists (and their lay enthusiasts) have championed biology as civilization's salvation. This narrative served both their professional interests and the ruling elites they explicitly denounced.[41] The professional managerial class simply has had no interest in revolutionary politics, since status quo arrangements largely benefit them.[42] Another *point de capiton* linking biomedicine and culture is the body. Bodies are artifacts of culture; as feminist scholars have shown, the body is always already gendered and raced.[43] Bodies do not shed their gender and race markings on entering the space of medical observation and intervention. There simply is no biology without these cultural signifiers, no way of "seeing" through them to abstract physiological principles. Ovaries are never *not* sexed; genes are never *not* raced.

The progression-ideology nexus raises not only political and theoretical questions, but methodological ones as well. How do science studies scholars explain the connections between dominant interests and institutional practices when the latter appear to be largely insulated and isolated from the former? How do these scholars study technological developments that seem, paradoxically, to both empower and discipline? More important, what method best comports with the assumptions I make about science, institutions, and culture—namely, that science is an institution that, while operating according to distinct logics (its cognitive core), nevertheless also manifests a particular political unconscious, one that is complicit with the dominant social and economic order? And that genomics, in particular, serves as the site for the recuperation of gender, race, and capitalism without the need for the explicit ideological appeals that characterized earlier periods of science's history?

To answer these questions, I turn to rhetorical methodology and a treatment of discourse as the transdisciplinary, institutional site for mapping the overlapping interests of genomics and society. Discourse is where institutional practices, cultural norms, and dominant beliefs converge. Researchers are embodied persons who must draw not only from an agreed-on and disciplinary-specific lexicon but also from the tropes, metaphors, narratives, and arguments that circulate outside of the scientific context but to which they are in no way immune. Moreover, the object of study— the body—is itself a discursive production, albeit presenting itself as a prediscursive, value-free materiality. Constituted by culture and dominant beliefs about race, sex, and sexuality, bodies in turn exert influence on the scientific discourses about them. Says Harding: "Scientific practices inevitably must 'deliver' nature to us in an already discursively encultured form."[44] This is especially the case in genomic medicine, in which the researcher's gaze shifts from the molecular to the somatic in an attempt to translate the products of genetic tests into cancer prevention. Scientific practices, then, will be shaped by—and will, in turn, shape—enculturated, normative bodies.[45] This interplay between discourse and inscribed material objects is what the philosopher Karen Barad calls the performative understanding of scientific practices.[46]

Discourse, then, is where habits of mind are reproduced, far removed from the intent or consciousness of its authors or from the material practices of the laboratory. More important, however, discourses traverse institutions and the larger social and economic structures of which they are part. Science may appear institutionally distinct from the larger culture, given the norms of objectivity and accompanying sanitized lexicon that are peculiar to it, yet because it is an institution, it comprises persons whose positionalities reflect an investment in certain class, race, and gendered relations. Specific disciplines, for instance, may have their own way of defining race, but it is nevertheless the case that certain hegemonic meanings of the term (that is, race as a biological artifact) emerge across the boundaries that define and otherwise separate them.

A rhetorical perspective attends not only to shared beliefs across mulitiple discourses but also to the inner workings of the texts that form them. How, for example, do certain meanings of the gene emerge? Or meanings of ovaries? Race? Risk? Put another way, how do objects become objects? In the field of science and technology studies, Jasanoff describes this as "the emergence and stabilization of new objects or phenomena: how people recognize them, name them, investigate them, and assign meaning to them; and how they mark them off from other existing entities, creating new languages in which to speak of them and new ways of visually representing them."[47] What Jasanoff captures in this passage is the way in which a rhetorical approach to scientific practices opens up space for considering change, but not in terms of a positivist history of science, replete with origins, linear unfolding of ideas, and presumptions regarding the ontological status of objects.[48] Rather, scientific practices—the conjunction of theories (and their assumptions about the material world), instruments, and observers—collectively enable the emergence of an object of study.

The rhetorical methodology I employ in this book not only explicates how this happens within the contingent, dynamic field of scientific knowledge production, but also contextualizes these changes within a historical context in which certain dominant interests have remained largely static.

Stabile's example of ultrasound is illustrative. In her essay on the emergence of the fetus as medical patient, she draws out the ways in which her theory of ideology helps us understand the connections between this new medical subject and the gendered, patriarchal order. She writes: "Where earlier appeals for 'motherhood' worked to erase female subjectivity and sexuality, this recent disarticulation functions not through ideology, but through the repression of material female bodies."[49] The fetus is not only a patient, it is a citizen subject, with rights, interests, and voice.[50] The newfound status of the fetus as subject relegates the woman to the status of mere container—a body in the service of another. Medical imaging technology thus serves two interests simultaneously: producing healthy babies and restricting women to passive spectatorship.

Women, therefore, are not necessarily hailed as mothers in order to serve institutional and state interests. In part, this is because explicit appeals to motherhood are less attractive and available options in an era marked by important political and economic gains for women.[51] Nevertheless, with the knowledge that those gains are threatening to dominant social and economic interests, one must consider how backlash and entrenchment are likely to occur. Medicine has historically served an important role in the disciplining of women in ways that have adapted to changing political and social realities; ultrasound, then, must be read as both a technology that visualizes the fetus and one that enforces politically and economically palatable gender norms and expectations. Although women have benefited from the disarticulation of the body from essentializing gender norms, in the medical context the emergence of the fetus as patient has consolidated medicine's control of women's bodies.

My treatment of the concept of genetic diversity in chapter 4 demonstrates the usefulness of rhetorically analyzing the constitution of scientific objects at the intersection of internal scientific practices and external cultural discourses of the body. It is socially and politically controversial to claim that race is biological; when geneticists do so, they are thoroughly, and very publicly, denounced.[52] Genomics itself, as it turns out, has provided the evidence for this denunciation, showing that most genetic diversity resides within, not between races, and that no allele has ever been found that is exclusive to one of the so-called races. But even though the concept of race as a biological construct is scientifically invalid, it is also tenacious—embedded in institutions and habits of language. It serves manifold interests in maintaining the status quo, from which science is hardly separate. A rhetorical analysis of genomics discourse shows that— despite the recalcitrance of the genetically diverse body that critics of race would say is a material reality that simply cannot be outmaneuvered by racialists—a racially coded body nevertheless persists, not in spite of the existence of genetic diversity, but in many ways because of it. My analysis shows, in fact, that relative genetic diversity now serves as the way ge- neticists can distinguish African Americans from other racial and ethnic groups. Genetic diversity, then, serves the interests of antiracialists and neoracialists alike.

The case of race thus shows the benefit of a rhetorical method that takes cultural studies' notion of hegemony as a starting point, insofar as dis- course is the means by which dominant ideas are both circulated as com- mon sense and shared among otherwise contrary interests. However, the language of these ideas is rarely, if ever, "the language of the rulers."[53] In the cases I examine, it is the language of science. This is the benefit of the rhetorician's unique ability to identify the relevant, enabling conditions of a specific institutional discourse (its material exigencies and constraints) in conjunction with its inner workings—the particular tropes, arguments, and linguistic arrangements expressed as scientific paradigms and experi- mental norms, together constituting its meaning-making practices.

Before I review the book's chapters, I must first explain its African American focus. I have already described the difference between under- standing race as lived experience and defining it as a biological state. I have also looked at this distinction, drawing on Stevens, according to the difference between a synchronic and diachronic understanding of race. Doing so helps us understand more clearly several phenomena in ge- nomics research and the application of that research in clinical medicine. These are the construction of the medical subject, specifically the intersec- tion of gender and race in surgical guidelines for oophorectomy (requir- ing an analysis of reproductive justice issues specific to African American women); the redefinition of disease from a product of racism to a prod- uct of evolutionary history; and the similar transfer of causal power from industrialization to genes in the field of environmental genomics. In all of these cases (the topics of chapters 3, 4, and 5, respectively), I proceed on the assumption that in order to attend to what is at stake in genomics' incursion into medicine and public health, we need both an examination

of racializing discourse and an explication of why this matters in a political and economic context marked by extreme disparities between African
Americans and other groups in income, employment, criminalization,
housing, and health. My analysis of this discourse depends on, and in
turn must reflect, the history of the race concept in the United States—in
particular, how it was employed first to define slaves as biologically distinct and thus inferior. Likewise, an analysis of the lived experience of
racism and how this is linked to observed health disparities must consider the specifics of racial oppression of African Americans in the US
context. Michael Brown and coauthors write: "The post–civil rights era
is also a world of high crime rates and joblessness in black communities,
with such deep enervating poverty that young, poor African Americans
are sometimes called the 'throwaway generation.' Many in the post–civil
rights generation see black students failing to graduate from high schools
and colleges at the same rate as white students, homeless black men and
women begging on the streets, jails full of young black men, and 'broken' black families."[54] These are phenomena that cannot and should not
be generalized to other groups. Indeed, because of the ongoing problem
of racial segregation, the unique characteristics of discrimination against
blacks, and the construction of whiteness in the US context as "not black,"
the "relationship of African Americans to whites therefore remains fundamental to any analysis of racial inequality."[55] In short, "the color of one's
skin still determines success or failure, poverty or affluence, illness or
health, prison or college."[56]

Chapter Overview

After the story I tell of genetics and hereditarian ideology in chapter 2,
chapter 3 begins the book's concentrated examination of three different
research foci in cancer genomics after the HGP. In that chapter, I locate
and describe an implicit reproductive politics in ovarian cancer genomics research. The rise of prophylactic oophorectomy as a thinkable, even
inescapable, surgery for women with BRCA mutations requires that a
gendered subject occupy this newly formed risk category. It is, as I show,
a subject position in which the female body is a reproductive one, since
the only acceptable path for resistance belongs to women who choose
motherhood, albeit at an early age. Genomics thus revives an ideology
of femininity that reduces ovaries to their reproductive function. But it
does so not in spite of feminist gains, but because of them. The work of

feminists in disarticulating the body from reproduction creates a space wherein it is thinkable to remove ovaries and breasts. The implicit assumption in genomics discourse is that to deny life-saving operations is to necessarily embrace an identity rooted in the normatively gendered body (as when preserving ovaries or breasts on the grounds that they are essential to feminine, sexual identity). Paradoxically, the conditions for the organs' removal are structured by traditional ideas of reproduction and motherhood: ovaries can be conserved, but only for a time, and only to answer to the desires of young motherhood. In contrast to those of white women, black women's ovaries are treated very differently: too often left in the body when cancerous, too easily removed to appease implicitly racist (and fabricated) anxieties over race and reproduction. And whereas young pregnancy and multiple pregnancies are permitted, even encouraged, in BRCA discourse (young pregnancy decreases cancer risk and fulfills procreative desires before genetic fate takes over), pregnancy has emerged in the cancer literature as a cancer risk for black women. Once again, black pregnancy is a pathological condition, a discursive development that must be read against the backdrop of a long-standing discourse in the United States linking black women, reproduction, and poverty.

In contrast to chapter 3, in which the BRCA subject is racially unmarked (thus masking white privilege), chapter 4 considers the context in which the BRCA subject *is* racially marked, that being the population turn in genomics research on breast cancer. The initial goal of BRCA sequencing may have been to serve the needs of women in what are called cancer syndrome families—those 5 to 10 percent of women with breast cancer whose disease is thought to be due to inherited genetic susceptibility—but it wasn't long before the prospect of wider applicability piqued the interests of researchers. Such a breakthrough occurred in 1995 with the description of three BRCA mutations found in women who were not related. It was simultaneously an "ethnic" turn, as the women with these three mutations all identified themselves as members of the Ashkenazim. With this development came the prospect that BRCA research (and in particular the BRCA test) would be relevant to a much larger group of American women: those connected through evolution, but genetically isolated through culture. Two years later, researchers published data gleaned from a study of African American women with breast cancer.[57] Chapter 4 argues that the population turn has simultaneously been a racial one. This is perhaps not surprising, since racial thinking was never fully rejected by the biomedical community, often operating surreptitiously under the cover of "population."[58] Nevertheless, this area of research warrants close investigation, as it not only raises the question of how racial thinking reveals itself amid emphatic denials by researchers that race has a biological basis, but also shows what that question means when situated within a larger, historically grounded understanding of the interplay between scientific knowledge claims and co-constituted institutional practices. The articulation of a racialized genetic subject—the black woman with or at risk for breast cancer—makes it increasingly unnecessary for us to target poverty and discrimination more generally in order to effect a reduction in health disparities and permit all communities to flourish. At times, this connection is made explicitly by geneticists, who insist that the lived experience of race and its connection to health disparities has been fully and adequately investigated. I make an additional, decidedly bolder claim than this and insist that the construction of race in genomics matters in a society that is, according to a number of indices, racist. Breast cancer risk becomes yet another means of pathologizing black bodies, in ways that not only affect the medical care and everyday life of African American women, but that also excuse a number of institutional practices and cultural discourses. The fact that the biological concept of race emerges within a literature diverse in its objects, research questions, fields, investigators, and language would suggest that certain habits of mind must surely be present, circulating and manifesting in the concrete practices of scientists. A sustained look at the race concept in a specific disease literature helps us understand not just whether the race concept exists, but how it persists, often unseen by the critical eye of either scientists or science studies scholars. Chapter 4 concludes with an examination of several projects in epidemiology that investigate how the lived experience of race and racism can account for the breast cancer disparities that BRCA researchers claim result from inherited mutations of genes.

After considering the specificities of racism and sexism as systems of oppression, in chapter 5 the book looks more closely at the direct links between women and industrialization, specifically by contrasting the gene-environment model (as envisioned by genomics) with theories of environmental breast cancer. I call the new science of gene-environment interaction "environmental genomics" and argue that it rests on politically invested assumptions about heredity, risk, and the body—namely, free-market ideology. Although environmental genomics by no means introduces such a logic into public health, it repackages it anew through a hereditarian-inflected rhetoric of progression, enabling the claim that we can rationalize environmental health interventions by identifying those who are truly at risk. Identifying and classifying persons at varying levels of hereditary susceptibility is not, however, progress as much as it is simply a different, albeit also value-laden, way of describing bodies, environments, and social relations. The discourse of environmental genomics presumes the individual to be the site of environmental health intervention—and only certain individuals at that. Two aspects of this discourse are particularly pertinent to my argument. First is the presumption that it is irrational, from a cost-benefit perspective, to regulate environmental pollutants without knowledge of individual genetic susceptibility, the implication of which is that current pro-industry regulations should remain largely unchanged. Second is the construction of a neoliberal subject of environmental genomics, through which behavior modification is seen as the common-sense course of action in response to toxic environmental exposures. The entrepreneurial consumer of new technologies and therapeutics becomes the agent moderating the impact of environmental exposures, thus leaving the market largely in control of what those exposures will be. This chapter ends with a gesture toward a different type of body politics, one grounded in embodied experience and social movement. In the book's conclusion, I extend this discussion by arguing for a postgenomics ethos, what I call a biosociality without genes.

2

Heredity as Ideology

Situating Genomics Historically

Two years ago I attended a panel discussion on epigenetics, part of the University of Georgia's "Darwin Days" series of events. I was interested in the panel because epigenetics exemplifies for me the stunning creativity and innovation of modern-day genomics research. Not surprisingly, Lamarckism came up—in particular, its association with Soviet-era genetics research, most of which was based on the materialist notion that environmentally induced changes could be transmitted to one's offspring. One of the panelists eventually proclaimed, "Well, I guess we're all Marxists now!" I took this comment to be an implicit recognition (even in its attempt at humor) that theories of gene-environment interaction are deeply political.

Yet despite the potential of gene-environment research to rethink the nature-nurture debate, in practice and in language it seems as though it is nature that always triumphs.[1] Even with the revelation announced not long before this book went to press that "junk" DNA, as it turns out, is no such thing, and may be even more important than genes, the gen*ome* remains central. In some ways, even epigenetics does not move us far afield from the centrality of heredity, as only those environmentally induced changes that can be inherited and that violate the implicit sanctity of the genome will be the drivers of research. Indeed, it is hard to fathom what sort of genetics research could displace the role that heredity firmly holds in the wake of the HGP and its extensive fanfare.[2]

What might be called an ontological complicity between genomics and the economic and social order has been a historical constant, even if its particular configurations have changed. It is essential to shed light on this complicity so as to correct for the mistaken belief that contemporary genomics bears no traces of the kind of thinking that informed earlier manifestations of hereditarian science. In this chapter I construct a historical narrative of genetics that draws on already published histories as well as original research. I begin with the eugenics movement of the progressive era, in which scientistic thinking enabled the belief that biology offered salvation from a range of social ills. Although much scholarly work has focused on how particular traits and behaviors (and the bodies that exhibited them) drew the interest and ire of eugenicists, I wish to draw attention to how eugenics discourse enabled a middle-class, professional subject position that was central to the salvation narrative of biologists. The chapter then examines posteugenics research, especially that which explicitly distanced itself from politics by engaging in laboratory experimentation with insects and other nonhuman life forms. The retreat to the laboratory, coupled with the contributions of scientists such as James Watson and Francis Crick (who, like many other key players, migrated from physics to biology, bringing with them an atomistic, molecular view of biology), permitted a crucial distancing of genetics from eugenics, the legacy of which was still felt in developments after World War II such as the rise of genetic counseling. Nevertheless, the molecularization of biology introduced a worldview that was no less political, no less invested in particular social and economic arrangements. Heredity became a fixture of public discourse, a triumph of nature over nurture, helping to reify the values most celebrated as alternatives to the evils of communism (and, by extension, the evils of socialism and collective resistance)—in particular, the ability of the individual to triumph within free-market democracy. In short, the hereditarian thinking of the postwar era represented a commitment to the status quo precisely at a time when it was most vulnerable to the challenges of rising social movements. The chapter concludes by situating the HGP within this social history of genetics, seeing it as much a product of the 1980s and the rise of neoliberalism as the fantasy of the so-called grail hunters. I argue the HGP and genomics more generally are implicitly committed to a paradigm of health and illness that is dependent on the conceptual transfer of causal power from the economic to the genetic, as well as on a particular gendered, racialized, and privatized social order.

The Eugenics Movement: Genetics and Ideology in Early-Twentieth-Century America

Eugenics, a term coined by the English scientist Francis Galton in 1883, is derived from the Greek *eugenes*, meaning good birth. Buoyed by renewed interest in Gregor Mendel's pea studies in 1900, the American eugenics movement was most active between 1900 and 1930.[3] Following popular family studies such as *The Jukes*,[4] Charles Davenport, a researcher at Cold Spring Harbor Laboratory[5] and one of the best-known American eugenicists, opened the Eugenics Record Office (ERO) in 1906 with funds from the family of the railroad magnate E. H. Harriman. The purpose of the ERO was to study the link between heredity and human behavior.[6] The American Eugenics Society was formed in 1923 and soon included twenty-eight state committees and a branch in southern California.[7]

Eugenicists such as Davenport believed that heredity explained phenomena like class, intelligence, criminal behavior, and disease—all problems perceived to be particularly acute at the beginning of the twentieth century. To provide support for the heredity theory, the ERO logged information gathered by its fieldworkers from house surveys, prison and hospital records, almshouses, and institutions for the deaf and blind as well as those classified as "mentally deficient" and "insane."[8] Davenport and his fellow eugenicists argued that the information they gathered demonstrated a link between biology and the pathological behavior they observed in the field.

For advocates of positive eugenics—the idea that genetically superior people should be encouraged to breed in order to propagate the best traits—ERO data were essential in making decisions about marriage and having children. Mendelism demonstrated that many traits were recessive and, if applied to human behavior, meant that family histories were essential to determine the chances of breeding those termed "undesirables." The "better baby contests" held throughout the United States are an example of the ways families were formally recognized for their enlightened breeding practices.

The notion of positive eugenics arguably explains why many progressives were also eugenicists,[9] in that they saw better breeding as an enlightened supplement to social welfare:

> The American [progressive] movement was in large part the creation of superintendents of asylums for the feebleminded, insane, and alcoholic, of prison wardens and prison physicians, of sociologists and social workers. They were in the forefront of the movement for bigger and better institutions to house and treat the unfortunate classes of mankind. They believed that society had a responsibility to care for the dependent and delinquent but that society had, at the same time, a responsibility to see that such persons did not contaminate the generations to come.

Despite the conservative implications of hereditarian thought, eugenics at first was closely related to the other reform movements of the Progressive Era and drew its early support from many of the same persons. It began as a scientific reform in an age of reform.[10]

Progressives embraced the general premise that the undesirables of society were not responsible for their condition—they were simply born that way. Such a view was not necessarily mutually exclusive with the idea that nurture was important. For progressives, the genetically unfit could still be—indeed, should be—improved by social services, although the larger goal was breeding out the traits that defined them.

After World War I, even this benevolent brand of eugenics—and the entire progressive movement—came under attack as critics argued that because of the immutability of genes, reforms of that era had been a waste of time. The war had resulted in a backlash against progressivism that had allowed eugenicist ideas to flourish. The cynicism of the 1920s was driven in part by a postwar wave of immigration of Southern and Eastern Europeans, viewed by nativists as a threat to American civilization and to the political order;[11] and in part by the highly publicized results of intelligence tests of army personnel, which had demonstrated that most members of the US armed forces were, by the tests' standards, of shockingly low intelligence.

For eugenicists, the postwar problems of poverty, crime, and social unrest demonstrated that the progressives' experiments with social welfare had been unqualified failures.[12] The eugenicists capitalized on the social and political climate of the time to champion negative eugenics, the idea that certain groups of people should be institutionalized, sterilized, or prohibited from immigrating to the United States. Mendelism demonstrated the necessity for harsh measures, since no amount of social change could counter the immutability of the "germ plasm."[13] Eugenicists went to great pains to explain the lesson of Mendel. For instance, in 1922 Albert Wiggam, a well-publicized eugenicist of the period, wrote:

> Educating you or cultivating your morals will never directly cause your children to be born brighter or more virtuous. If your father went crazy from getting hit on the head with a brickbat, you do not inherit his cracked brain, but only his inability to dodge brickbats. Stupidity begets stupidity, and intelligence begets brains; but a thousand years of educating or improving the parents will never "improve" the children. In short, "Wooden legs are not inherited, but wooden heads are."[14]

What was at stake in the nature-nurture debate of the 1920s was no less than civilization itself—a rhetorical shift that characterized the backlash of the decade. Wiggam elaborated in mid-decade:

> Charity, hygiene, sanitation, and the triumphs of chemical, biological, and medical genius have preserved for reproduction great numbers of incompetents and defectives who in the good old days of natural selection would have died young. Civilization is thus seen to be a self-destructive enterprise, the most dangerous enterprise upon which mortals ever set out, and one which, by its very nature and objective, sets going those agencies which in the end destroy the men that built it and insures its own dissolution.[15]

Circulating this discourse of "race" suicide, eugenicists successfully lobbied for some of the most reactionary legislation in US history. In 1924 the Immigration Act was passed by overwhelming majorities in both the House and the Senate and was signed by President Calvin Coolidge. For the next three years, immigration was limited to a small percentage of those nationalities recorded in the 1890 census, in an effort to target the Eastern and Southern Europeans blamed for race deterioration and social unrest (since there were far fewer Eastern and Southern Europeans in the United States in 1890, the immigration restriction affected them most severely). And in 1927 the Supreme Court upheld the legality of some twenty state sterilization laws in *Buck v. Bell*.[16]

Stakeholders in contemporary debates about genetics point to the eugenics movement as evidence of the power of genes to justify extreme, misguided discrimination. Yet it is also important to consider the extent to which eugenics included an unstated politics that was palatable to diverse audiences. As I pointed out above, many eugenicists were progressives, illustrating the appeal of scientism for professionals who saw themselves as the experts most able to address the effects of industrial expansion. In fact, the middle class provided the bulk of support for the eugenicist cause. The membership in the American Eugenics Society never exceeded 1,200 members, but the middle class was well represented. The movement's leaders belonged to the professional class, including physicians, social workers, clerics, writers, and professors in the biological and social sciences.[17] And although the group's membership was small, eugenic theories reached large audiences. The ERO devised many methods for disseminating its research to a largely middle-class audience. According to Daniel Kevles, a revolving audience of attendees came to Cold Spring Harbor to discuss this research, information about it was given to legislative committees, bulletins and books were published regularly, and lecturers were dispatched in response to numerous requests for speakers. More importantly, fieldworkers from the ERO often became teachers of genetics and eugenics policy or became members of "state commissions and other institutions dedicated to the reduction of hereditary degeneration and defect."[18] By 1924 there were 250 fieldworkers and 750,000 cards of data. Eugenicists like Albert Wiggam also reached large audiences of middle-class readers through general interest magazines.[19]

The successful appeal to the professional and managerial class, in fact, allowed eugenics to play an important role in maintaining the legitimacy of the era's power structure.[20] However, biologizing social problems was not only about blaming the victims of industry; in fact, criticism of the upper class was a prominent theme in eugenicist discourse at the time. The professional and managerial class was thus situated as the objective, scientific mediator between the upper class and its victims—with both groups providing evidence of what can occur as a result of bad breeding.

To the professional and managerial class, the scientism of eugenics was expressly positioned as an alternative to politics. From the point of view of the middle class, conditions after World War I suggested the inadvisability of radical politics of any kind, whether that of the ruling class or labor—especially militant, socialist labor groups. Not only should citizens concern themselves with "the economic fear that our industrial civilization has overreached itself and is carrying about in its body the seeds of its own decay," Glenn Frank—the editor of the *Century*, a general-interest magazine pitched to middle-class audiences—told his readers, they should also fear that "the crowd-man and crowd-processes of thinking will take the place of the creative and insurgent individual who has hitherto been the mainstay of progress."[21]

This sentiment squared nicely with efforts of eugenicists to forge a new ethos grounded in an apolitical yet benevolent individualism. Harvey Wiley, a eugenicist writing in *Good Housekeeping*, described the ethos of the "new American" as follows: "He will not be a doctrinaire. He will not be devising, in the seclusion of his study, a new millennium. He will be intensely practical and not theoretical. He will look upon his fellow men not as objects of exploitation, but of collaboration. He will finally rise a free and independent individual, above the fetters of collectivity. He will be more than a mere number in a herd. He will have individuality and initiative."[22] To be practical and independent meant studying and living by the laws of heredity, not subscribing to any particular ideology, on the Left or the Right. Lothrop Stoddard wrote in *Revolt against Civilization*, a eugenicist tract, that Bolshevism violated the natural hereditary order: "Against this formidable adversary stands biology, the champion of the new."[23] Lewis Terman, who led the successful effort to institutionalize intelligence tests, went so far as to suggest that "laborers," who are not as genetically fit as "managers," would "drift easily into the ranks of the anti-social or join the army of Bolshevik discontents" if not "trained" properly.[24] Indeed, the genes of leftist leaders themselves were called into question: in the proposed immigration restriction legislation of 1914, "revolutionaries" were categorized under "psychopathic constitutional inferiority," essentially a eugenic category.[25] Thus the turn to nature, a turn for which the science of genetics was particularly well suited, came at the expense of questioning, in any meaningful way, the status quo.

Reform Eugenics: Geneticists Retreat to the Laboratory

After achieving most of its political successes during the 1920s, the "mainline"[26] American eugenics movement experienced a steady decline during the 1930s.[27] During this decade, the rise of fascism in Europe and the economic crisis drove many scientists to the Left,[28] which arguably accounts for what historians characterize as eugenics' decline in popularity among the general public. The Third International Congress of Eugenicists, held in New York City in 1932, drew fewer than a hundred attendees.[29] By 1940 the ERO, headed at that time by the zealot Harry Laughlin, closed its doors.

In an effort to respond to critics, several prominent scientists openly condemned the views of mainline eugenics and articulated what historians have called "reform eugenics." Reform eugenicists "rejected in varying degrees the social biases of their mainline predecessors yet remained convinced that human improvement would better proceed with—for some, would likely not proceed without—the deployment of genetic knowledge."[30] Specifically, reform eugenicists discarded mainline views concerning race:

> In the reform-eugenic view, society needed the reproductive contribution of all competent people. Mainline concern with "the race" was beginning to be replaced by attentiveness to "the population." The new

language was more than just a change in terminology; it reflected the reform eugenicists' belief that valuable characteristics were to be found in most social groups, and that the best in human variation was to be encouraged.[31]

Reform eugenicists included H. J. Muller, a student of the well-known fruit-fly researcher Thomas Hunt Morgan, who would receive considerable attention for his work inducing mutations in *Drosophila*, and later, for his theories concerning the mutagenic effects of radiation.[32] The American Eugenics Society, which became the primary locus of eugenics activity after the demise of the ERO, also represented this kinder, gentler version of eugenics, especially during the leadership of Frederick Osborn.[33]

Reform eugenicists advocated the development of a human genetics research program. This would explore more methodologically sound ways of explaining the mechanisms of heredity, in part by focusing on traits that could be practically explored as Mendelian—such as eye color and blood type—which in turn would lay the foundation for linkage analysis.[34] In 1937, for instance, researchers observed a link between hemophilia and color blindness on the X chromosome.[35] For eugenicists, such developments carried important practical implications: if disease could be linked to particular regions of chromosomes (by their association with Mendelian traits already located there), presumably individuals (and couples) bearing those Mendelian traits could be counseled as to the probability of bearing children with isolated diseases.[36] This dovetailed with efforts by reform eugenicists to focus less on behavior and more on medical pathology.

Indeed, a crucial aspect of the reform eugenics agenda included the growth and promotion of genetic counseling. One of the earliest genetic counseling clinics was the Heredity Clinic at the University of Michigan, which opened in 1940 and was headed from 1946 to 1981 by James Neel, an MD and PhD noted for his genetics research in Japan after World War II.[37] By the 1950s, counselors could employ biochemical tests[38] to determine whether a parent (or potential parent) carried a recessive gene.[39] For a few diseases, they could also offer predictions as to whether additional children would be born with the same disorder as their sibling. In other cases, counselors used family medical histories to estimate the risk of having children with various diseases or disorders. Although actual demand for genetic counseling was low,[40] it was a service actively promoted to US audiences.

Nevertheless, key revenue streams for molecular biology, including research on the transmission of heritable traits and the exact physical structure and chemical composition of genes, bore traces of the eugenics legacy, if only indirectly. Newly formed molecular biology programs drew much of their funding from private foundations that remained keenly interested in the relationship between genetics and human behavior (the term "molecular biology" was in fact first coined in 1938 by Warren Weaver, a Rockefeller Foundation trustee).[41]

The rising popularity of molecular genetics had much to do with the recognition in the 1930s that eugenics was a liability—it was bound to racist, nativist, and reactionary views and policies, which in turn were based on bad science. Geneticists wished to distance themselves from both. For example, when Morgan publicly attacked eugenicists in 1925 for confusing nature with nurture and ignoring more "enlightened" explanations for human behavior,[42] he represented many plant and animal geneticists who distanced their work from both eugenics and the emerging field of human genetics. Many researchers who worked during the 1930s recalled being discouraged from having any association with human genetics on account of its origins in the eugenics movement. As Kevles notes, "indeed, in the United States, plant and animal geneticists tended to discourage prospective colleagues from having anything to do with human genetics, reminding them that it was associated with the racism, sterilizations, and the scientific poppycock of mainline eugenics."[43]

The Rockefeller Foundation, an earlier supporter of eugenics, likewise distanced itself from the field in the 1930s, although it never lost interest entirely in the relationships among biology, heredity, and behavior (that is, social pathology). The foundation's goal in the 1930s was to recuperate the basic science via well-funded interdisciplinary programs, bringing together scholars from a variety of fields and being very careful not to publicly link the basic science to any particular agenda. But, as Lily Kay's history of the foundation reveals, it envisioned molecular biology as a salvation science, much like its eugenicist predecessors: "The paucity of knowledge about physical mechanisms of gene action represented merely a delay in understanding the basis of intervention; lack of knowledge meant a need for fundamental research. The implicit belief in unit characters,[44] even among life scientists, persisted well into the 1940s, along with the intuitive expectations of eugenic intervention."[45]

Specifically, the Rockefeller Foundation justified its support of molecular biology on the grounds that it would "rationalize" the human sciences and provide a scientific basis for social control, a fundamentally eugenicist desire.[46] Funding for molecular biology was part of the foundation's Science of Man project, a project "articulated in terms of the contemporary technocratic discourse of human engineering," the goal of which was to "develop the human sciences as a comprehensive explanatory and applied framework of social control grounded in the natural, medical, and social sciences." It was, moreover, "erected on the bedrock of the physical sciences in order to rigorously explain and eventually control the fundamental mechanisms governing human behavior, placing a particularly strong emphasis on heredity."[47]

Like eugenics, then, molecular biology promised to explain social phenomena—in particular, social phenomena that threatened the status quo and the particular sets of social relations unique to it. As Kay put it, molecular biology was part of the foundation's larger goal of "restructuring human relations in congruence with the social framework of industrial capitalism."[48] Indeed, the Great Depression in many ways shaped the foundation's programs and explains their intentions:

> The trauma of the Depression catalyzed complex national reactions, unleashing widespread soul-searching and reappraisals of the work ethic, private enterprise, and the equation of prosperity with biological and social fitness. True, blacks and the unskilled lost their jobs first, with whites and managerial personnel being the last to go; joblessness and shiftlessness ranked highest at the lowest echelons. Eventually, though, the Depression claimed a toll among the poor and rich, the slothful and industrious, the godless and the devout. The cloak of benevolence fell from big business, stimulating bolder flirtations with communism and socialism. The stained hand of business tainted its spiritual ally, the Protestant churches. The downward spiral in church attendance paralleled a shift to the Left, and many church leaders and ministers vented their anticapitalist sentiments from the pulpit and in print.[49]

In response to this growing anticapitalist sentiment, the Rockefeller Foundation assumed responsibility for producing subjects who would restore the image of the market—namely, by effectively, and happily, internalizing its core values. The Science of Man project was designed to maximize what the foundation considered desirable traits through the study and manipulation of all aspects of physiology, including genetics. Informed by the appropriate biological knowledge, the human sciences could then help promote the breeding of persons who were mentally stable, physically fit, resistant to disease, and in control of their so-called drives.[50] Social control was already a dominant paradigm in the human sciences, on which the foundation could draw for its vision for molecular biology: "A little reflection will show that all social problems are ultimately problems of social control—capital and labor, prostitution, taxes, crimes, international relations."[51] The foundation thus put into place an early version of sociobiology, but with an explicit focus on serving the needs of the capitalist order. Molecular biology served this goal by providing the scientific means for producing good, dependable workers (similar to eugenics, molecular biology could easily tie pathologies such as mental illness to political resistance to the economic order). Moreover, the scientific management of human behavior could serve as a justification for scaling back social programs. No more useless funds would be wasted on the unfit. If the political and social culture of the 1930s had brought nurture into the foreground, the Rockefeller Foundation would counter with an intellectual and material investment in nature.

The Great Depression had also, according to Kay, sown the seeds of a backlash against scientists and engineers; the excesses and failures of industrial capitalism (now under even more scrutiny because of the fantastic failures of the market) were seen as inextricably tied to the expertise behind their so-called rationality. Thus, the Rockefeller Foundation's foray into the biological sciences would also help rehabilitate the image of the scientist expert. As in the eugenics era, scientists could once again be seen as enlightened, nonpartisan experts serving the interests of the public good. Indeed, "the launching of the Rockefeller Foundation's new deal for biology as part of its commitment to the human sciences could not have come at a more auspicious time."[52]

The DNA Revolution and Postwar America

The molecular biology research supported by the Rockefeller Foundation in the 1940s augmented the field of molecular genetics, with many researchers focusing their investigations on the relationship between the three-dimensional structure of molecules and their functionality.[53] For James Watson, Francis Crick, and Rosalind Franklin, the molecule generating the most interest was deoxyribonucleic acid (DNA), shown to be the "agent" of heredity in 1944.[54] In April 1953, Watson and Crick published an explanation of just how this agency was exercised (they did not acknowledge Franklin's indispensable contribution)—specifically, the theory that the DNA molecule was structured as a double helix, with the two helical chains held together by nucleic acids, usually referred to as "bases."[55] A month later, Watson and Crick explained what they termed the "genetical implications" of the double helix theory by describing how DNA transmits a "code" when the helix separates and how messenger RNA replicates the base sequence.[56] This process, they argued, "may help to solve one of the fundamental biological problems—the molecular basis of the template needed for genetic replication."[57]

This research would eventually be said to usher in a revolution in biology. Indeed, terms such as the "DNA revolution" and the "molecular revolution" became ubiquitous in discourse about genetics research, both popular and scholarly. Yet when Watson and Crick first announced the theory of DNA structure and function to the scientific community, there was little public attention to it. Not until many external events—social, economic, and cultural—had unfolded would the language of revolution be used, which suggests, following Thomas Kuhn, that scientific revolutions are possible only when enabling social conditions are present.[58] In what follows, I draw out the many ways in which the 1950s were an important decade in the history of genetics, its institutional interests, and its ideological investments. During this period, geneticists relied less and less on foundation funding as human genetics became more mainstream; heredity became a significant part of public culture, largely through intense national attention to the dangers of nuclear testing; Cold War culture inextricably bound DNA to the principles of individualism and the free market; and molecular biology blossomed as a science of life, further extricating genetics (rhetorically, anyway) from the realm of the social and political (whereas eugenics, reform eugenics, and sociobiology were essentially social programs, molecular biology would be a scientific program with social implications).[59] Indeed, although this period was marked by a sharp demarcation between genetics and nature, on the one hand, and human sciences and nurture, on the other, the postwar period would lay the foundation for the eventual triumph of the former in the 1980s, most notably with the HGP. The 1950s was also a socially and politically significant decade insofar as science and ideology were newly distinguished from each other, after the line dividing them had become muddied not only during the eugenics era but also in World War II, with the atrocities associated with the US use of nuclear weapons. As Evelyn Fox Keller has shown, the history of molecular biology is not complete without attention to the significant number of physicists who turned their attention to biology, seeking refuge in the "secrets of life" after their stigmatizing association with the "secrets of death."[60]

What were the important developments that contributed to the rise of genetics—in particular, of DNA as the agent of heredity—as a culturally significant scientific discourse in the 1950s? One factor was the considerable increase in federal funding for genetics, especially human genetics. Despite advances made in this area in the 1930s and 1940s, efforts to establish human genetics programs at universities failed to attract much interest before World War II.[61] After the war, however, scientific interest in human genetics increased greatly. The American Society of Human Genetics was formed in 1948,[62] and in 1954 the *American Journal of Human Genetics* published its first issue.[63] By 1959 the society had approximately 500 members: "In the fifties, human genetics in the United States attracted a number of new recruits, both PhDs and, increasingly, MDs, aided and abetted by the opportunities for study and research available because of the government's interest."[64] Also important for the field were many advances in medical genetics. For example, in 1948 Neel showed sickle cell anemia to be linked to a recessive gene. Chromosomal studies also furthered the field significantly. In 1959 researchers found that trisomy 21 (also known as Down syndrome) is caused by the production of three copies of chromosome twenty-one instead of the typical two copies. Physicians and medical researchers were also increasingly drawn to human genetics because of postwar advances in biochemical and molecular genetics, especially research involving enzyme deficiencies (as with inborn errors of metabolism) and studies of microorganisms and drug resistance.[65] After the war, genetics was gradually added to medical school curricula and, in many cases, independent departments of medical genetics were created.[66] The change was dramatic: although a "little over half the medical schools in this country and Canada in 1953 offered some instruction in genetics and only seven offered full courses in medical genetics," by 1972 courses in genetics were required in half of all American medical schools and offered in all but a quarter.[67]

Federal funding for human genetics research also spiked in the 1950s because atomic age interest in the mutagenic effects of radiation became a national security concern. Thus, one funding stream for genetics was the Atomic Bomb Casualty Commission, headed by Neel,[68] whose goal was to study the molecular effects of nuclear fallout. Researchers studied the children of women in Japan who had been exposed to radiation while pregnant, although data was also collected in order to further basic research in heredity.[69]

Perhaps more significant than the money spent on radiation research was the fact that the science was very much a public affair. Fears about radiation-induced illness from nuclear bomb tests, X-rays, and occupational exposures—exacerbated considerably by extensive media coverage—rendered heredity a national concern. Geneticists openly voiced fears about reproductive damage from excessive radiation exposure. Some scientists appropriated a familiar crisis rhetoric of the time by invoking the image of a "race of monsters."[70] In 1954, for example, the New York Times, referring to an article by Alfred Sturtevant, a researcher at the California Institute of Technology, wrote: "The human race will reap a macabre harvest of 'defective individuals' because of the increase in 'background radiation' caused by only the few atomic and hydrogen bombs that have been exploded to date."[71] "The entire human race," the article continued, "is being subjected to the risk from radiation."[72] The impact was thus defined as global in scope and devastating to the future of civilization: according to Sturtevant, "the effects of successive exposures are cumulative. They are permanent in the descendants of the affected genes. There is no recovery."[73] More important, perhaps, is evidence that human genetics had not fully parted ways with its eugenicist predecessors. The Times article included the claim by Sturtevant that genes are crucial to all human development, including "mental characteristics."[74]

In 1956 the National Academy of Sciences released a report by its Genetics Committee on the biological effects of atomic radiation, ensuring that the issue of heredity would remain a matter of public significance. The report was funded by the Rockefeller Foundation (Warren Weaver of the foundation was head of the Genetics Committee), and its full text was published in the New York Times on June 13. The title of the accompanying front-page story—"Scientists Term Radiation a Peril to Future of Man; Even Small Dose Can Prove Harmful to Descendants of Victim, Report States"[75]—reflected the media's focus on the part of the report in which the authors claimed that "for the general population, and in the long run, a little radiation to a lot of people is as harmful as a lot of radiation to a few, since the total number of mutant genes can be the same in the two cases."[76] That same week a New York Times editorial invited comparisons between peacetime radiation pollution and nuclear war:

What the geneticists on the committee of the National Academy predict is a slow, almost imperceptible deterioration of the human race, a
deterioration that may take centuries. It is a deterioration marked by a
higher death rate, a lower birth rate, a lowered resistance to disease, a
proneness to leukemia. Unless this deterioration is halted or controlled
the end is just as certain as if atomic bombs were to destroy every community on earth.[77]

Articles like these no doubt contributed to making heredity a significant part of public consciousness. Nevertheless, apocalyptic claims about
"race suicide" were refuted by many scientists and by results of the Atomic
Bomb Casualty Commission demonstrating that the "only" serious effect of the US bombing of Japan was a significant increase in leukemia
cases. In other words, no "frightening stories of strange after-effects of
the atomic bombing" had been documented.[78] And Muller, the geneticist-
cum-public intellectual and born again conservative, argued that fears
over a race of monsters were generated primarily by communists as a way
to stir up public opposition to US nuclear testing.[79] Although the long-
term effects of radioactive fallout were significant, Muller argued, to halt
all tests and atomic research more generally would amount to a Cold War
victory for the Soviet Union.

In the event, the National Academy of Sciences report did not recommend an immediate halt to nuclear weapons tests. It did not even recommend a halt to the use of X-rays in medicine, the area of use the report's
authors concluded was the greatest cause for concern. Rather, the report
recommended that exposure be monitored, measured, limited if possible,
and, most important, studied. Thus genetics, and especially that part of
the field concerned with human DNA, had been given a mandate of unprecedented importance. This mandate did not include questioning the
importance of nuclear science and the atomic age it had helped to create, but it did make genetics research as important as physics, a science
that had considerable cultural capital largely because of its ability to define universal laws of the material world. Genetics promised to unlock the
secret laws of the life world—a point that might have become particularly
important because more and more physicists were crossing disciplinary
lines to study biological systems, especially after World War II.[80] These
disciplinary crossings also marked another important moment in the history of science and its relation to society: the retreat to the science of life
would afford a safe haven from politics and ideology for physicists who

had become unavoidably mired in Cold War politics, largely as the targets of antinuclear activism.

By the early 1960s, the idea of unraveling the genetic "code" was generating considerable excitement in the media, and the language of "life" was characteristic of the discourse in part because of the language of scientists themselves.[81] And, more often than not, molecular biology was compared to physics. In 1958 *Time* published a short piece titled "The Secret of Life," which described genetics as "the comparatively young, fast-developing science of heredity that is trying to solve the mystery of life, as physics works at solving the mystery of matter."[82] Dramatically elevated to the level of space exploration, human genetics had now become a frontier to be conquered: "Our scientists are now close to a feat that will rival the conquest of space—the unlocking of the secrets of heredity, reproduction, and of life itself."[83] For some commentators, DNA research surpassed all other scientific endeavors in importance:

> Indeed, it is safe to say that some of the biological "bombs" that are likely to explode before long as a result of [breaking the code] will rival even the atomic variety in their meaning for man. Some of them might be: Determining the basis of thought. . . . Developing remedies for afflictions that today are uncurable, such as cancer and many of the tragic inherited disorders.[84]

"There is no doubt," this writer concluded, "that the breaking of the genetic code will rank among the greatest scientific achievements of all time."[85] When the Nobel Prize in Medicine and Physiology went to Watson, Crick, and Maurice Wilkins in 1962,[86] molecular biology had become nothing short of a revolutionary endeavor:

> The unveiling of the chemical structure of DNA has opened the way to the ultimate understanding of the chemistry of life and the mechanism of heredity. Through it, the science of biology has reached a new frontier, said to be leading to "a revolution far greater in its potential significance than the atomic or hydrogen bomb."[87]

The importance placed on a single molecule was driven in part by the significance that scientists like Crick imputed to it. In 1958 Crick published his "central dogma" theory.[88] According to Crick, the central dogma stipulates that DNA is the primary source of genetic information:

DNA transfers information to RNA, which is then used to make proteins.[89] Further elaborating the point in 1970, Crick specified that there are some "special transfers" of genetic information (for example, the transfer of genetic material from RNA to DNA in retroviruses), but there are "3 transfers which the central dogma postulates never occur": protein to protein; protein to DNA; and protein to RNA.[90] Moreover, he argued, "the discovery of just one type of present day cell which could carry out any of the 3 unknown transfers would shake the whole intellectual basis of molecular biology, and it is for this reason that the central dogma is as important today as when it was first proposed."[91] The central dogma imputed tremendous significance to the molecule that Watson and Crick had described with so much interest and passion. In part, this reflects, again, the desire of scientists like them to place biology on par with physics—to rationalize the field by proposing a simple, and more importantly, a (nearly) universal law governing the transmission of genetic information.

Moreover, the central dogma was a direct refutation of any theory of acquired characteristics. According to Robert Olby, "the claim that the machinery for the transmission of this chemical specificity from DNA to proteins only allows it to operate in the one direction constituted the molecular basis for the rejection of the possibility of any form of Lamarckian heredity."[92] The central dogma was largely interpreted to mean that there is only so much genetic potential that the individual can realize, regardless of environmental context. Richard Levins and Richard Lewontin argue that the central dogma was in many ways an ideological throwback to the Weismann-inspired belief of the immortality of the germ plasm.[93] But this throwback was also a means of distinguishing American genetics from Soviet genetics[94]—put another way, it distinguished the central dogma from the revolutionary tenet of Lamarckism that "theories which deny the reality of change are generally associated with loyalty to the status quo."[95]

To understand how the DNA revolution represented loyalty to the status quo, it is important to consider the economic, social, and political context of the 1950s.[96] Indeed, this decade was a time of significant transition. On the one hand, the economic prosperity of the time made it possible to continue supporting, and in some cases expand (for example, Social Security) the social welfare programs of the 1930s and 1940s. On the other hand, poverty remained a significant problem, especially for women, people of color, the elderly, and those lacking a formal education. Social reforms fell far behind social change, as when women's wartime employment opportunities did not extend to the postwar period despite their desire to remain in the workforce. The middle class expanded considerably during this time, the key beneficiary of the economic largesse of the period. The laboring class benefited as well, although labor lost political ground and much of what was understood to be economic security was anything but: for this class as well as the middle class, economic status was possible because of significant federal subsidies, housing construction being just one example. And the expansion of the professional and managerial class[97] mattered a great deal culturally, socially, and politically. It signaled a significant expansion of a consumerist, entrepreneurial ethic; it also expanded the population for whom a centrist, noncontroversial politics would be attractive. The Eisenhower administration in particular represented this kind of orientation:

> The Eisenhower administration proved singularly successful in offering a respite from conflict and controversy, charting a path so close to the middle of the road that it was difficult for critics to gain an audience. For Ike, moderate Republicanism meant running the government in a businesslike fashion, giving power back to private interests in the states where appropriate, but retaining the new overall responsibility of the federal government for issues of social welfare and security. As he described it, the government should be liberal when it comes to human beings, conservative when it comes to spending; and if—as Adlai Stevenson said—that meant recommending more schools "to accommodate the needs of our children, but not [providing] the money," the rhetoric nevertheless sounded right to millions of Americans tired of new departures. Even Democrats seemed to go along. As Joseph Rauh, the liberal ADA attorney, declared in 1956, "Congressional Democrats have become practically indistinguishable from the party they allegedly oppose." The Eisenhower legislative program pleased pro-business conservatives, while not being sufficiently reactionary to alienate moderates and liberals.[98]

This was a politics broadened and magnified as a result of the Cold War: free will, individualism, nonconformity (think of the Cold War–era revulsion against the so-called organization man), and the market were all rhetorically deployed as alternatives to orthodoxy, state power, and communism. Politics and activism on the Left in the United States were increasingly looked on with suspicion and, by extension, so were social

welfare programs and the role of the state in ensuring the economic security of US inhabitants. Says the historian William Chafe: "In the end,
the 1950s represent more a time of transition than of stolidity. During the
immediate aftermath of World War II, the possibilities for massive social
change in the condition of workers, blacks, and women had been snuffed
out. Instead, the appeal of the consumer culture, suburbanization, and a
respite from political conflict became dominant, deflecting attention toward achievement of the material goods necessary for the 'good life.'"[99]

This particular narrative of the 1950s helps us understand just what science studies scholars mean by the triumph of nature over nurture that the
DNA revolution brought about. Keller, for example, writes:

> Of signal importance in the transfiguration of genetic determinism is
> the fact that, in the late 1960s, molecular biologists began to develop
> techniques by which they themselves could manipulate the "Mas-
> ter Molecule." They learned how to sequence it, how to synthesize it,
> and how to alter it. Out of molecular biology emerged a technologi-
> cal know-how that decisively altered our historical sense of the immu-
> tability of "nature." Where the traditional view had been that "nature"
> spelled destiny and "nurture" freedom, now the roles appeared to be
> reversed. The technological innovations of molecular biology invited
> a vastly extended discursive prowess, encouraging the notion that we
> could more readily control the former than the latter—not simply as a
> long-term goal but as an immediate prospect. This notion, though far
> in excess of the actual capabilities of molecular biology of that time,
> transformed the very terms of the nature-nurture debate; eventually, it
> would transform the terms of molecular biology as well.[100]

"In short," Keller concludes, "in the vision inspired by the successes of
molecular biology, 'nature' became newly malleable, perhaps infinitely
so; certainly it was vastly more malleable than anyone had ever imagined
'nurture' to be."[101]

The 1950s would lay the groundwork for the near permanent triumph
of nature over nurture. Indeed, by the time molecular biology had obtained both the public attention and material resources it needed to be a
scientific program in its own right, it was clear that the nation's creativity, its sources of innovation, and, in short, its economic power, would be
harnessed to explore the basis of life—of nature—and how to control it.
Nurture would increasingly become attached to our Cold War enemy and

to the leftists and other activists here at home. The radical atomization of biology, of life itself, would find a comfortable home in a culture marked by a celebration of the individual over the collective. All of these features would also help the triumph of the free market—if nurture was best represented by redistribution of wealth, nature represented a far less costly, and therefore politically and culturally acceptable, way of solving the problems associated with poverty, low wages, and economic insecurity.

Children of the Revolution

The major premise of the molecular revolution would not be borne out in practice; in fact, considerable scientific evidence now undermines the agency of DNA. Scientists have learned that the development of organisms is quite complex and cannot be reduced to a genetic code.[102] As Lewontin puts it,

> elements in the genetic messages may have meaning, or they may be periphrastic. The code sequence GTAAGT is sometimes read by the cell as an instruction to insert the amino acids *valine* and *serine* in a protein, but sometimes it signals a place where the cell machinery is to cut up and edit the message; and sometimes it may be only a spacer, like the periphrastic "do," that keeps other parts of the message an appropriate distance from each other.[103]

This explains, for instance, why the sides of the same fly can develop differently even though the same genes are present and the fly is exposed to the same extracellular environment.[104] Yet despite knowledge of the ways in which biological development is much more a dialectical systems affair than a unidirectional, hierarchical unfolding of a DNA code's instructions, in the early 1990s the United States began a project to sequence all three billion base pairs of the human genome.[105] The National Institutes of Health (NIH) and the Department of Energy (DOE) together spent $2.7 billion dollars on the project until 2000, when its completion was announced, and that figure does not include the money spent by the Sanger Institute in Cambridge and around the world.[106]

Historians date the origins of the HGP to the mid-1980s,[107] and its full-scale launch in 1991 represents the convergence of various professional, economic, and cultural interests: hereditarians who remained committed to the genetic basis of all human disease and behavior; grail hunters devoted to embarking on a genomics project unprecedented in scale and scope; Cold War laboratory interests seeking a new mandate for the nation's defense infrastructure; and a biotech industry that saw early on that gene sequences and the biological material used to study them were a potential source of capital it could accumulate and exchange as a result of large-scale, very fast sequencing. Moreover, several influential biomedical scientists were interested in improving mapping and sequencing technology since interest in the role of genes in disease, such as cancer, was growing considerably. Technology existed at the time to map and sequence individual genes of interest,[108] but the process was labor- and time-intensive. A large sequence database, it was argued, would result in faster, more efficient ways of finding and studying genes of interest and would allow scientists to explore what, if any, importance other areas of the genome possibly held.

One of the first meetings about the project was convened in 1985 at the University of California, Santa Cruz, by then-chancellor Robert Sinsheimer, a long-time molecular biology enthusiast who was interested in a project that would bring prestige to the relatively young campus.[109] It was also in 1985 that Charles DeLisi (trained in physics and formerly chief of mathematical biology at the NIH), then head of the DOE's Office of Health and Environmental Research, proposed human genome sequencing as a post–Cold War project for the DOE's extensive infrastructure of high-tech laboratories capable of both rapid ("high-throughput") sequencing and high-volume data storage. The DOE, long interested in genetics research (in particular, the relationship between radiation and mutation), had established the "Genbank" database in 1983 at Los Alamos National Laboratory, which could hold large quantities of sequence information.[110] In 1986 DeLisi held his own workshop, similar to the one in Santa Cruz, to discuss the technical aspects of a genome project.[111]

Also in 1986 advocates of a large-scale sequencing project held a meeting at Cold Spring Harbor to discuss its merits.[112] As in the 1950s, genetics, this time in the form of the HGP, was described in fantastical terms, likened to previous projects that had advanced scientific understanding and technological prowess in ways unforeseen when they began. Advocates also drew on the legacy of Cold War discourse, imputing to the HGP profound ontological and existential attributes, thus elevating science beyond the political and ideological fray. In a 1986 editorial in *Science*, Renato Dulbecco, a Nobel laureate in medicine, not only argued that a full sequence was a prerequisite for further progress in cancer genetics research but likened such an effort to the space program.[113] A year later, Walter Gilbert described a full sequence as a biological "Rosetta Stone" that once completed would fulfill the human imperative to "know thyself."[114]

Soon after the Santa Cruz, Sante Fe, and Cold Spring Harbor meetings, DeLisi allocated money from the 1987 DOE budget for a five-year genome project, with support of the department's secretary.[115] DeLisi also had important congressional support, as Republican Senator Pietro "Pete" Domenici of New Mexico (home of the Los Alamos National Laboratory) saw the genome project as a way to revitalize support for the national laboratories. At congressional hearings held by Domenici, advocates included scientists touting the possible medical benefits of a human genome project and industry representatives arguing that it would help the United States compete economically with Japan and Europe, where human genome projects were being enthusiastically discussed.[116] Thus the HGP was seen early on as an important part of the US strategy to compete in an increasingly global economy.

With the DOE pushing ahead, it was only a matter of time before the NIH staked its claim, thereby ensuring that the genome project would move forward and with considerable support. Uneasiness at the idea of a DOE genome project was widespread due to the department's history of health research abuses;[117] the NIH, moreover, could claim jurisdiction (and would be compelled to) simply because of the biomedical implications of the research.[118] Eventually Congress would approve funds for fiscal year 1988 for both the DOE and NIH, although the NIH would get substantially more money and would serve the role of lead agency.[119]

Concern over DOE involvement was by no means the only argument made against a human genome project. Nevertheless, debate was confined primarily to the technical sphere[120]—critics argued that the HGP was technologically impractical, that it would siphon off funds from other areas of the biosciences,[121] that biomedical research could proceed effectively without it,[122] and that a large project of this nature would negatively impact professional life.[123]

These arguments proved easy to address. The National Research Council (NRC) issued a favorable report of the project in February of 1988 and, notably, the report's authors included scientists originally opposed to a genome project.[124] The report's recommendations were clearly meant to address the skepticism of biomedical scientists: it advised that new money be found for the project and that work begin with physical and genetic mapping, thereby accelerating the search for specific disease-related genes (as opposed to sequencing all of the genome data first, without any regard to whether or not particular regions were important for biomedical research). This was "a type of research that many biologists wanted to pursue anyway."[125] After the release of the NRC report, James B. Wyngaarden, director of the NIH, called a meeting chaired by David Baltimore, an early skeptic (the meeting also included other critics, like David Botstein): "There, Baltimore, Botstein, Watson, and the rest of the advisory group closed ranks behind a project that would proceed along the lines recommended by the NRC committee's report."[126] In October of 1988, James Watson agreed to head the project.[127]

The fact that prestigious scientists, many of whom had been highly critical of the HGP early on, came to endorse it helped shore up its scientific credibility considerably. Still, a project that would make the genome so central to biomedical science and, by extension, medical practice, would raise a number of ethical and policy concerns. Questions over employment and health discrimination, informed consent, and the potential psychological harms of genetic testing fueled much of this debate. These concerns were largely preempted by Watson's announcement in December of 1988 that the NIH would set aside 3 percent of its HGP funds for the study of the project's ethical, legal, and social implications—what would become known as the ELSI program. Although unexpected, Watson's announcement was not entirely surprising since he had witnessed firsthand what can happen if geneticists appear reckless and unregulated. For in the early 1970s, scientists had successfully transplanted, for the first time, genetic material from one organism to another using restriction enzymes.[128] Since recombinant DNA research involved experimentation with potentially dangerous viruses, there were obvious safety concerns, and some scientists publicly voiced them.[129] The publicity resulted in research moratoriums and, more worrisome for the research community, proposals to ban the research outright.[130] Watson's decision was no doubt a shrewd one; ELSI funding would "give all the main would-be alarmists—the most likely candidates for the money, after all—a substantial interest in keeping the source of the money (the genome project) alive."[131]

Thus, the history of the HGP reflects the culmination of many converging institutional interests, interests that prevailed because their opponents' arguments were limited to the technical sphere; because social discourse was virtually nonexistent, thanks to the privileged status of reasoning in the technical sphere; and because the ELSI program had been unilaterally established by Watson. What I would also like to consider is how and why the HGP may be read as a product of the social and political culture of the 1980s, all the while bearing traces of eugenicist ideology.

Indeed, much evidence suggests that the project was explicitly linked to a eugenicist-style program of social reform, albeit in the reform eugenics tradition of the progressive era and later of the Rockefeller Foundation. HGP enthusiasts envisioned this program as a way to ameliorate many of the social problems of the 1980s, problems that they saw as ultimately rooted in the transmission of defective genes. Daniel Koshland, the editor of *Science*, sounded very much like a progressive-era eugenicist when he wrote in 1989:

> The benefits to science of the genome project are clear. Illnesses such as manic depression, Alzheimer's, schizophrenia, and heart disease are probably all multigenic and even more difficult to unravel than cystic fibrosis. Yet those diseases are at the root of many current societal problems. The cost of mental illness, the difficult civil liberties problems they cause, the pain to the individual, all cry out for an early solution that involves prevention, not caretaking. To continue the current warehousing or neglect of these people, many of whom are in the ranks of the homeless, is the equivalent of providing iron lungs to polio victims at the expense of working on a vaccine.[132]

Similar sentiments were expressed by the authors of an Office of Technology Assessment report published in 1988: "one of the strongest arguments for supporting human genome projects is that they will provide knowledge about the determinants of the human condition" and that they will illuminate the causes of diseases that are, in turn, the basis of many "societal problems."[133]

Juxtaposing Koshland's and the report's comments with the breathtaking cutbacks in social welfare spending that marked the 1980s puts this particular version of hereditarianism in perspective. In 1981 Congress cut more than $25 billion from welfare programs, agreed to cut hundreds of billions more, and cut taxes over five years by $750 billion. President Ronald Reagan won congressional approval for a $1.2 trillion increase in defense spending over the same time period.[134] Affirmative action was aggressively fought by the US attorney general, the Civil Rights Commission, and the Equal Employment Opportunity Commission. Food stamp benefits were cut, as were Aid to Families with Dependent Children benefits, resulting in a large reduction in the number of persons receiving welfare—and for those still eligible, benefits were cut considerably.[135] The Reagan administration also reversed many of the significant environmental gains of the 1970s (a topic I explore in chapter 5). These cutbacks would result, among other things, in increased homelessness, especially among the mentally ill—the very people that Koshland and others claimed would be best served by a large-scale genome project. Indeed, the Reagan years are perhaps as well known among progressives for the massive increase in homelessness among the mentally ill as they are for the "Great Communicator's" ability to portray a social and economic landscape that would supposedly bring increased prosperity to all. Reagan, relying principally on Office of Management and Budget Director David Stockman and the latter's embrace of the neoliberal economist Arthur Laffer, maintained that "the American spirit of individualism, competition, and personal pride would be restored, and with the shackles of government bureaucracy removed individual citizens would once again be liberated to maximize their abilities and aspirations."[136]

The Reagan years would be especially hard on people of color, especially African Americans. The climate of the time was characterized by vicious attacks on blacks and by a concomitant rise in biological theories of racial pathology. Racist biologism was not limited to the likes of Richard Hernnstein and Charles Murray and their infamous (and thoroughly discredited) *Bell Curve*, published in 1994. By the late 1980s, the notion that violence associated with African Americans was a biological (that is, racial) trait was firmly in place, leading in 1992 to the NIH-funded Violence Initiative.[137]

It is within this context of significant cutbacks in the social safety net, deepening poverty, a widening rich-poor gap, and a rise in vicious and racially motivated public and scientific discourse that the largest attempt ever to make heredity foundational to the human and biological sciences was made. As in previous moments in genetics' history, biology—especially that focused on inherited defect[138]—served as a salvation science, offering a rational, apolitical, and objective approach to social problems. In fact, within this salvation narrative, social problems ceased to be social at all; they were problems of human (and sometimes racial) evolution. The material devastation of the neoliberalism of the 1980s—homelessness, poverty, and environmental pollution—could now be subsumed within the discourses and practices of science. Moreover, that part of the professional and managerial class made up of salvation scientists would expand considerably, the combined effect of billions of dollars in federal money, significant expenditures in the private sector, and the widespread (and savvy) public presence of geneticists like Francis Collins.

More than solving social problems, the HGP would help solve crises of capitalism itself. Indeed, the generation of a new form of capital—biocapital[139]—would make the so-called science of life more than that. It would produce value, thus opening up possibilities for new kinds of profit and wealth. And it would valorize market values by making the body practices of biomedicine key sites for the production and consumption of biomedicine itself. The concept of biocapital captures the ways in which market mechanisms overdetermine scientific practices, as well as how the convergence of market speculation, commodification, and new epistemologies have ushered in a new and unique phase in the history of capitalism more broadly. What has been central to this formation is the development of technologies for sequencing and storing large amounts of genomic information. Genomics companies patent sequences in the hopes of selling access to the information to pharmaceutical companies or to conduct their own pharmaceutical research. What Kaushik Sunder Rajan calls "promissory futures" fuels speculation around the vast amounts of information that is generated and stored. What has made such speculation possible is the way that genomics has decoupled biological information from material biological sources (cells and tissue), thus permitting the circulation and exchange of the former.[140]

I would add to Sunder Rajan's analysis that biocapital also reconstitutes, and indeed strengthens, the heretofore indirect and figurative investment that scientists, as members of the professional and managerial class, have in the social and economic order. More important, it is an investment more obscured than ever by a rhetoric of health and healing. The HGP era has been conspicuously marked by considerable expansion of the notion of genetic disease, a conceptual (and, Keller says, ideological) development that began in the 1970s but really secured its place in the scientific and public imaginary with the launch of the HGP. Advocates of the project, Keller says, not only expanded "genetic disease" to include virtually all diseases for which a genomic component could be hypothesized (which is a considerable number) but also tied the project to a right to health—in particular, a right to a normal, healthy genome. "Health," she says, displaced "culture" and "perfection" in the language of genetics.[141] Keller observes that with the genome project the grand narratives of the eugenics era have been replaced by grand narratives both of the existential significance of the genome project and of its ability to cure all disease, broadly defined. As I explained in chapter 1, disease is both the embodiment of social relations and the means which the analysis and critique of those relations is permanently deferred. Through the trope of health, genomics, as an institution, "explains the world in such a way as to make that world appear legitimate."[142]

To examine just how this new health discourse legitimates the status quo, I turn to the BRCA (short for "breast cancer") genes, since they became an intense object of study at the same time that the HGP was wielding its influence in the world of genetics and in public discourse more broadly. BRCA research and related projects help us understand the connection between genetic disease and the needs of both capital and the state, mediated through practices that call into being the medical subject—one that both inhabits a normatively coded body and consumes what Sunder Rajan calls the excess surplus value of the genetic test. The biological substance of biocapital, however, is both gendered and racialized; indeed, biocapital moves through bodies as well as databases and laboratory media. Indeed, the concepts of disease and health are no less ideological than those of culture and perfection.

BRCA and Genetic Disease in the Post-HGP Era

In the year the HGP officially came to a close, Evelyn Fox Keller wrote that genes "have carried us to the edge of a new era in biology, one that holds out the promise of even more astonishing advances. But these very advances will necessitate the introduction of other concepts, other terms, and other ways of thinking about biological organization, thereby inevitably loosening the grip that genes have had on the imagination of the life scientists these many decades." For Keller, the concepts "distributed program" and "developmental stability" are perhaps suitable alternatives to DNA-centered notions of "program" and "genetic stability":

> We have long known that the rate of protein synthesis requires cellular regulation, but now we have learned that even the question of what kind of proteins are to be synthesized is, in part, answered by the kind of state of the cell in which DNA finds itself. In higher organisms DNA sequence does not automatically translate into a sequence of amino acids, nor does it, by itself, suffice for telling us just which proteins will be produced in any given cell or at any stage of development. Like

the responsibility for maintaining fidelity, this work too is distributed among the many players involved in post-transcriptional regulation. The same can be said regarding the determination of how a protein functions.[143]

Still, one might argue that these contingencies do not deny the fundamental role of DNA in encoding for certain biological processes, even if the translation process acts more like a dispersed network and less as a hierarchical informational system. However, significant evidence denies even this level of agency for DNA. According to Keller, we need to think in terms of a "distributed program in which all the various DNA, RNA, and protein components function alternatively as instructions and as data."[144] We now know, for example, that during "alternative splicing," "spliceosomes" (a type of protein) combine with five small molecules of RNA to cut apart segments of messenger RNA. According to Barry Commoner, this "can be said to generate *new* genetic material," since in alternate splicing, "the gene's original nucleotide sequence is split into fragments that are then recombined in different ways to encode a multiplicity of proteins, each of them different in their amino acid sequence from each other and from the sequence that the original gene, if left intact, would encode."[145] Researchers have also discovered prions, proteins that code for other proteins but do not comprise amino acids, thus violating the central dogma postulate that there can be no protein-to-protein transfers.[146] Moreover, chaperone proteins ensure the proper folding of other proteins[147] and, thus, "play an *active role* in setting the limits of fidelity" (emphasis added).[148] Despite these developments, it appears that genes have maintained their grip on the scientific and popular imagination. Those genes include the BRCA genes.

One of the first cancers to be included within the heritability paradigm was breast cancer. In the 1940s, Eldon Gardner studied family pedigrees in Utah that appeared to lend credence to the idea that breast cancer susceptibility can be passed on to offspring.[149] Indeed, the evidence for the heredity hypothesis is compelling: some families are marked by multiple cases of breast and ovarian cancer (some women being afflicted with both), early age of onset (women getting breast cancer in their twenties and thirties), frequent occurrence of bilateral cancer (cancer in both breasts), and breast cancer among male family members. The theory behind these observations is the so-called two-hit hypothesis:[150] if BRCA genes are tumor suppressor genes when working properly, then mutations of both copies of one of these genes provide the conditions for cancerous tumors to develop. Those women who inherit one bad copy of either gene at birth (the first hit) are at greater risk for this outcome—if their one good copy mutates and becomes nonfunctional due to endogenous or exogenous carcinogens or the vagaries of cellular replication (the second hit), cancer develops.[151]

In the 1980s, Mary-Claire King, a geneticist at the University of Washington, began investigating the link between DNA and breast cancer by studying family pedigrees and performing linkage analysis—a technique greatly enabled by new technologies developed throughout that decade.[152] Then in 1990 Mark Hall, King, and their colleagues published their hypothesis in *Science* that a "breast cancer gene" called BRCA1, on the short arm of chromosome 17, was linked to breast cancer.[153] Although only 5–10 percent of breast cancers were estimated to be of the heritable kind, King and her group hypothesized that finding the BRCA gene could help explain all breast cancer, thus broadening the impact of this research at a time when breast cancer was becoming a significant public issue.[154]

After the 1990 study was published, other labs in the United States and abroad, including those involved in the HGP, began the process of determining the exact location and chemical makeup of the so-called breast cancer gene. In 1994 the sequence of BRCA1 was published, and it has been confirmed that its transmission follows an autosomal dominant trait pattern.[155] In 1995 another breast cancer susceptibility gene, BRCA2, was sequenced. And, in 1991, soon after the first report linking BRCA and breast cancer, Steven Narod and colleagues published the finding that ovarian cancer risk was also linked to mutations of BRCA1; it is now known that mutations of BRCA2 are also implicated in ovarian cancer incidence.[156]

Mutations of the BRCA genes are considered high penetrance:[157] women who inherit a mutated copy of one of the genes carry a risk for breast or ovarian cancer far greater than women who have not. Nevertheless, the mutations are relatively rare, thought to be present in one out of every 300–500 people, although researchers have described mutations of these genes that are more common to discrete populations—for instance, persons of Ashkenazi descent (a topic explored more in chapter 4). In fact, a 2005 task force report recommended against routine testing of the general population, although newly updated guidelines from the American Congress of Obstetricians and Gynecologists recommend routine screening for a fairly broad segment of women.[158]

As of 2010 approximately 3,500 variations of BRCA1 and BRCA2 have been reported to the Breast Information Core database, about half of which have been categorized (that is, placed in a category of suspected mutation; not all variations of the BRCA genes have been implicated in tumor development).[159] In addition to the focus on the BRCA genes, as well as other genes thought to explain the 5–10 percent of inherited breast and ovarian cancer cases, geneticists have turned their attention to common polymorphisms[160] of genes that increase risk for cancer—not nearly to the extent of high-penetrance genes like BRCA, but that are alleles present in much larger groups of women.[161] These inquiries are not as advanced as BRCA research, but they nevertheless mark a crucial expansion of the genomics paradigm, described by Francis Collins and coauthors as no less than the "reclassification of all human illnesses on the basis of detailed molecular characterization" and "the introduction of a new molecular taxonomy of illness."[162]

BRCA tests have been commercially available since 1994, despite reservations among members of the research community at the time. Several companies and clinics offered the test until Myriad Genetics legally won exclusive rights to it after successfully securing and defending patents on both the BRCA1 and BRCA2 genes.[163] Indeed, the BRCA genes have been a fruitful source of capital for Myriad Genetics (providing funds for its various research and development programs), which has also patented its BRCA test, "BRCAnalaysis." (The biocapital landscape is quickly changing, as both Myriad's patents are due to expire and are being challenged in court as this book goes to press.)[164]

The transformation of genetic material into biocapital could not have occurred were it not for media discourse about the BRCA genes that helped construct a narrative of danger, heroic discovery, and empowerment. In many ways, public discourse about BRCA mirrored that about the HGP insofar as the main characters were geneticists, hunting for and finding disease genes in the face of considerable odds. Yet there were important differences, not least of which is that whereas the language of the HGP was mostly limited to rather grandiose, metaphysical talk about coming to "know ourselves," the BRCA genes enabled a more concrete discourse of health, providing the means for solidifying, once and for all, the shift from cultural perfection to health at the intersection of genetics and society.[165]

Richard Lewontin, describing the mystifying discourses of science, writes:

For an institution to explain the world so as to make the world legitimate, it must possess several features. First, the institution as a whole
must appear to derive from sources outside of ordinary human social
struggle. It must not seem to be the creation of political, economic, or
social forces, but to descend into society from a supra-human source.
Second, the ideas, pronouncements, rules, and results of the institution's activity must have a validity and a transcendent truth that goes
beyond any possibility of human compromise or human error. Its explanations and pronouncements must seem true in an absolute sense
and to derive somehow from an absolute source. They must be true
for all time and all place. And finally, the institution must have a certain mystical and veiled quality so that its innermost operation is not
completely transparent to everyone. It must have an esoteric language,
which needs to be explained to the ordinary person by those who are
especially knowledgeable and who can intervene between everyday life
and mysterious sources of understanding and knowledge.[166]

One way that the science of genomics secured these attributes was
through media discourses of the early 1990s. Genomics became a significant part of the public consciousness largely because of extensive
mass-media coverage of the BRCA genes. Between 1990 and 1994, over
300 articles about the genes were published in US newspapers alone;
this number does not include the many articles published about other
genes believed to play a role in breast cancer.[167] Inherited breast cancer
research was a regular topic for major papers like the *Boston Globe*, *New
York Times*, and *Washington Post*, as well as popular news magazines
like *Newsweek*, which featured cover stories about BRCA1.[168] This research was newsworthy first and foremost because the isolation and description of BRCA1 and BRCA2 represented a significant achievement
for the genetics community. Moreover, it made real one of the promises
of the HGP—that it would provide new weapons in the fight against
cancer, one of the most talked about and feared diseases.[169] BRCA was
at times characterized as a "deadly" gene,[170] and the very high risk associated with its mutations helped construct a discourse of fear. For at the
time that BRCA research entered the public spotlight, it was known that
mutations of the gene could increase a woman's risk for breast cancer by
as much as 85 percent over the course of her lifetime. Coupled with this
fact was the belief that many more women harbored these mutations
than previously thought. The president of the American Cancer Society

claimed in 1994 that "inherited gene mutations may play a role in 5 per-cent of all breast cancers and 25 percent of cases that occur before age 30. But breast cancer is so common that the mutations might be found in 1 in 200 women—or about 600,000 in the USA."[171] The "1 in 200" figure was widely quoted in major media;[172] indeed, two years later, it appeared in a *Boston Globe* report in which the reporters asserted—de-spite evidence at the time that mutation risks varied, especially when family history is taken into account—that all of these women carried highly pathological mutations.[173] The supposed high frequency of muta-tions led one journalist to essentially rewrite the history of the research by claiming that "the commonness of the mutation in the general popu-lation explains why the gene was so hungrily anticipated and its discov-ery so celebrated."[174]

BRCA's presumed relevance for the fight against all breast cancer also contributed to its newsworthiness. By the time that BRCA research began in earnest, breast cancer had become a significant public issue. In 1992 President Clinton committed an unprecedented amount of money to breast cancer research, and politicians and corporations discovered breast cancer as a campaign issue and public relations opportunity.[175] Indeed, most of the early BRCA stories made mention of the research's presumed potential to explain all breast cancer, and when early studies lent credence to this claim,[176] the media paid even closer attention.[177]

The search for the BRCA genes was particularly well suited for wide-spread coverage because the key players were well-known geneticists, members of a discipline with elite status. Moreover, the possibility of map-ping a gene for a common cancer captured the imagination of the genet-ics community itself—until this point, only those genes associated with relatively rare, Mendelian (single-gene) diseases had been located. Thus, when King first presented her team's evidence that strongly suggested a link between BRCA1 and a region of chromosome 17, the genetics commu-nity took notice.[178] According to one researcher, "very few genes have riv-eted the biomedical research community and the public as has BRCA1."[179] Subsequent histories and recollections of BRCA research would narrate the mapping of BRCA as the outcome of visionary researchers involved in nothing less than a paradigm shift in genetics research. In 1996, two years after the gene had been found, Kevin Davies, the editor of the pres-tigious journal *Nature Genetics*, and Michael White, a science journalist, published a popular history of this research, *Breakthrough: The Race to Find the Breast Cancer Gene*. They characterized their book as the story behind BRCA1—the people that defied the skeptical research community to chart its location amid the vast oceans of the human genetic blueprint, or DNA, the people that devised the intricate techniques to manipulate our DNA in order to isolate the gene, and the people who led the furious international race to be the first to discover this precious morsel of DNA that is BRCA1.[180]

Reflecting on advances made by 1996, the geneticist Stephen Friend wrote that "with similar feelings as to when readers of George Orwell's classic futuristic book, *1984*, finally lived through the calendar year of its title, we are already passing through a time only imagined less than a decade ago when the first human breast cancer susceptibility gene was identified."[181] The imagery of BRCA research as a race among geneticists all over the world fit within the conflict frame the media tends to rely on and thereby helped circumvent the difficult task of reporting on a highly complex science.[182] All the major papers, for example, appropriated this imagery of BRCA research as a contest to find the gene: *USA Today* proclaimed in one article that "the race to find the first gene linked with inherited breast cancer is over"[183] and in another that "after a four-year hunt, scientists have identified a gene that causes inherited breast and ovarian cancer. The gene's existence was discovered in 1990 and since then laboratories worldwide have raced to locate the Breast Cancer 1 (BRCA1) gene."[184] The *Los Angeles Times* employed similar imagery: "The race to locate the gene was ultimately won by a team of more than 45 scientists headed by geneticist Mark H. Skolnick of the University of Utah and Myriad."[185] The *Atlanta Journal-Constitution* referred to the research as a "decade-long treasure hunt and the beginning of a new era in cancer research."[186] Related to this, media discourse also fetishized the difficulty of genetic mapping, a not uncommon practice since genomics reporting began in the mid-1980s.[187] BRCA researchers, for example, were often called "gene hunters,"[188] participating in the "frenzied sifting of 600,000 letters of genetic code on chromosome 17 where the gene was believed to be located."[189]

Although locating the BRCA genes *was* difficult, given the sequencing and mapping technology available to geneticists at the time, the media's nearly sole focus on the competitive nature of BRCA research and the sheer difficulty of finding a breast cancer gene enacted the rhetorical effect wherein describing the gene became an end in itself—a phenomenon with precedent in the public discourse about the HGP. The social history of this science was effectively erased. More than that, an implicit genetic reductionism was at work, since finding the BRCA gene was confused with understanding its function and relation to breast cancer—the act of mapping (wrongly) implied clinical significance to the larger public. In the event, mapping the gene was just the first step in the long, unpredictable process of understanding, and treating, a complex disease. Nevertheless, what was initially just a matter of a technological breakthrough in linkage analysis became a regular part of health and medical discourse for women with, or at risk for, breast cancer. Preventative measures, such as taking tamoxifen and undergoing preventive mastectomy, quickly achieved the status of common sense.[190]

At the same time that the BRCA genes were garnering so much media attention, alternative explanations for rising breast cancer rates were being vehemently discredited or outright dismissed. In 1994, just one year after a major study attempted to recontextualize breast cancer as a public health issue by hypothesizing a link between chemical exposure and breast cancer risk, the media pronounced the debate about environmental breast cancer all but over—despite the fact that some subsequent studies confirmed this link.[191] A *Los Angeles Times* headline of April 20, 1994, "Breast Cancer Study Clears DDT, PCBs," was characteristic of media discussion at the time.[192] (In 2002—when the main study of the Long Island Breast Cancer Study Project was released, showing no connection between pesticides and other chemicals with breast cancer risk—naysayers, including the *New York Times*,[193] used the results to discount not only the connection between pollution and breast cancer on Long Island but also any research pursuing the environmental hypothesis. The editors of the *New York Times* wrote that "these negative findings suggest it is time to rein in this fruitless quest.")[194]

Overall, environmental breast cancer research was depicted in the mainstream media as politically motivated, the product of activists' scare tactics. Dominant representations portrayed them not as educated and informed residents who understood the scientific challenges of environmental epidemiology,[195] but as well-meaning political neophytes "pestering" lawmakers to act.[196] "Advocates" "felt" like there was a connection between pollution and breast cancer; "experts" "knew" there was not.[197] All in all, activists were portrayed as eager bystanders, merely "happy," "delighted," and "thrilled" that scientists were finally paying attention to the environment[198] and responding to their fears (residents were routinely depicted as easily "alarmed" by research showing the relationship between various cancer clusters and pollution sites). The theory of environmental breast cancer was characterized as the product of a faith-based movement driven by the tragedy of breast cancer, not a legitimate, albeit evolving, scientific hypothesis.

As activists began to wield considerable political influence, representations in the media shifted accordingly. By 1998 the image of the Long Island activist had shifted from political newcomer to seasoned lobbyist and fundraiser. But rarely were the activists presented in a positive light. One of the most detailed stories to appear in the *New York Times* on the issue of breast cancer on Long Island did not deal with the science behind the Long Island studies but with the internal politics of breast cancer activism there. Its paternalistic-sounding title, "Growth Pains on Breast Cancer," was followed by the observation that the movement "has evolved from a simple hands-on volunteer operation into sophisticated charitable enterprises that are increasingly being run like businesses."[199] The naïve activist is characterized as inevitably falling victim to the corrupting influences of money and power. Exhaustive details regarding how much money was raised—and how—and the political lobbying activities of group members were dominant themes. Shifts in alliances among members, including the launch of new organizations, were portrayed as problematic; unnamed "critics" and "many activists" were quoted saying that problems like "bitterness and resentment" were associated with the "power struggle to maintain their identity and receive credit for their accomplishments."[200] The fact that some people quit their jobs at various breast cancer organizations was assumed—without the benefit of elaboration or evidence—to be suspect and not, simply, the normal course of social movement evolution.[201] My point here is not that the media misrepresented breast cancer activists, only that implicit in this coverage is the view that the environmental breast cancer hypothesis was inherently unscientific. In contrast, genomics was depicted as scientifically credible—which is to say, value- and interest-free.

And lastly, also influencing the widespread acceptance of genomic medicine in the 1990s, was the unprecedented growth of federally funded bioethics. Paradoxically, BRCA testing became part of routine medical care amid considerable public debate about its ethical, legal, and social implications. To make sense of this apparent incongruity, one must examine more closely the fundamental assumptions of bioethics as well as the conditions of its emergence in public discourse about BRCA.[202]

Attention to the ethical, legal, and social implications of genetics results from two powerful stakeholder interests in genomics: industry and science. In regard to the latter, geneticists have been keenly aware of the controversial nature of genetic testing. In the year when the BRCA test first became commercially available, Francis Collins, then the director of the National Human Genome Research Institute, addressed an audience in San Francisco:

> If we do this wrong . . . the consequences could be a public outcry . . . from which we will not recover for decades. . . Genetic discoveries and genetic testing will not be dangerous additions to medicine. They will be very valuable once we understand enough about them. But one fear is that we will commit enough egregious errors . . . that the American public will be totally turned off and will decide . . . that they don't want anything to do with genetic technology. . . We don't need a genetic thalidomide.[203]

As of 2003, the federal ELSI program had funded at least sixty-seven studies on the topic of BRCA testing. Not one had interrogated the legitimacy of the test itself, thereby ensuring that the social history of BRCA research (and, indeed, of genomics more generally) would not occupy its proper place in public debate.[204] This outcome is largely due to the fact that institutional discourses, such as those produced by the media, the government, and bioethicists, rely on a liberal humanist perspective on science and society. By this I mean a perspective that understands scientists—driven by the accumulation of intuitively compelling facts about the material world—to be the agents of scientific change whose work may (indeed must) be interpreted outside of particular historical and material contexts. An ahistorical, instrumentalist view of science inheres in this view, for one of its central tenets holds that the conditions of scientific work do not overdetermine subsequent applications. Science is neither inherently good nor inherently bad; rules and norms simply need to be developed to ensure its ethical appropriation by medical practitioners, employers, and the like. These rules and norms typically reflect yet another component of the liberal humanist worldview: that the principal right to be protected is the right to choose freely whether one wants to participate in the genetics revolution. This is typically the perspective of the scientists, ethicists, and individual women whose public testimony shape the narratives of media discourse. And it is the perspective that receives the most funding of the multi-million-dollar federal ELSI program, the largest bioethics research program in history.

In short, the dominant narrative about the ethics of BRCA testing presumes that predictive diagnostics are powerful tools that provide lifesaving information for many women (even if the treatment options aren't perfect), but that along with the hope and benefits of scientific revolutions come costs like insurance and employment discrimination and psychosocial distress. The best way to ameliorate these costs, according to this narrative, is to educate physicians, institutionalize genetic counseling, and protect the privacy of women. With protections like these in place, more women can reap the benefits of the genomics revolution—and in some cases help it along. For example, in an essay in the *Baltimore Sun* (a piece picked up by several other major papers), the legal theorist Karen Rothenberg argued that when women choose to not get tested (attributed by her to fears of employment or insurance discrimination), scientific research is the principal casualty. Regarding a specific instance, she wrote: "Understandably concerned about discrimination against herself and her children, [the woman] refused to participate in the long-term follow-up studies that could help medical science learn more about prevention, early diagnosis and treatment of familial breast and ovarian cancers. Nor will her gun-shy family participate in any more research studies, even though the benefits to future generations could be enormous."[205] Indeed, Rothenberg describes genomic medicine in such a way as to make it unthinkable that any reasonable person would refuse, provided the requisite legal protections are in place to protect insurance converge, employment, and privacy. In the same piece in which she discusses the specific case of a woman refusing to be a research subject, Rothenberg begins with a reminder that "genetic testing is indeed a scientific miracle. Never before in history have we had the amazing ability we have today, to test for genetic predisposition to common killers such as breast or ovarian cancer and heart disease. And new genetic tests for other dread diseases are appearing almost every day."[206]

Public discourse about the BRCA genes—the collective result of sensationalizing genomic medicine, silencing environmentalists, and the unprecedented growth of a particular bioethics perspective—meant that the only real question addressed was how best to integrate genomics into routine medical practice.[207] Yet, as the next chapter shows, what risk means, as well as who occupies the epistemological and ontological category of those at risk, has undergone a significant shift. BRCA medicine has developed in such a way that women with mutations are patients in much the same way as women with cancer. What sort of body practices have emerged as a result?

3

Genomics and the Reproductive Body

In 2008 National Public Radio interviewed Jessica Queller, author of the book *Pretty Is What Changes*. In the book and the interview, Queller describes the experience of testing positive for a BRCA mutation, one that conferred upon her an 87 percent chance of being diagnosed with breast cancer and a 44 percent chance of an ovarian cancer diagnosis. These percentages presented themselves as a form of terrifying, imposing yet ultimately empowering knowledge. For Queller this meant deciding between one of two courses of action: heightened vigilance in the form of routine screening or prophylactice surgery to remove her breasts and ovaries so that she could, as she put it, defy her destiny. And, as the title of Queller's book suggests, contemplating surgery of this kind is tantamount to radically questioning hegemonic notions of femininity and beauty.

At the time of the National Public Radio interview, Queller, then thirty-five, had already undergone a double mastectomy; she had decided to postpone having her ovaries removed. The reason for this deferral? She desired marriage and children, so the decision to undergo an oophorectomy would have to wait—even though her mother had died of ovarian, not breast, cancer. We see, then, in Queller's narrative how gender norms are reestablished in the process of becoming a patient. As Queller said in a later interview, referring to her breast reconstruction surgery, "you're really put back together again beautifully."[1]

Queller's story, as it turns out, represents the most recent twist in an evolving discourse of gendered medicine. For women at BRCA-related risk for breast cancer, oophorectomy, the prophylactic removal of ovaries, helps them reduce that risk while avoiding the visible scars of mastectomy. Yet at the same time, plastic surgery techniques have advanced enough so that mastectomy is becoming more popular with both women who test positive for the BRCA mutation and those diagnosed with breast cancer.[2] Moreover, women like Queller are advised to undergo oophorectomy by age thirty-five. She refused to accept this deadline, choosing cancer if need be in order to exercise reproductive autonomy. Decisions like Queller's remain, for the most part, unquestioned. This chapter explores why.

Specifically, the chapter both maps conceptual shifts whereby the removal of organs for prophylactic purposes becomes thinkable and contextualizes those shifts within structuring social relations that are themselves both gendered and racialized. Heredity has come to function as an object of medical diagnosis and intervention. Heredity, then, effectively materializes as disease—a rhetorical phenomenon. Hereditarianism informs the very knowledge claims in genomics discourse that make possible the *treatment* of BRCA status—in some cases, a treatment that is more radical than for women actually diagnosed with cancer. Risk, in other words, has become actionable because genetic risk is taken to be substantively and materially different from family history (and hence constitutes a novel scientific object).[3]

As genomics has moved from the laboratory to the doctor's office, the field of vision has shifted accordingly. Specifically, the object of discourse is the whole of the corporeal body, itself inscribed with cultural norms. These norms account for both the symbolic and material recalcitrance of the body of genomics. Indeed, mining the research on ovarian cancer offers an opportunity for exploring a particularly striking example of the link between inherited risk and the body and for examining how genomics discourse both appropriates and rehabilitates social values, even while engaging in what would be considered sound research practices. Bodies are necessarily resistant to medical technology—oophorectomy has to be carried out in particular way, within certain time constraints, and women who undergo this procedure must endure additional medical interventions (for example, hormone replacement therapy). But even these seemingly natural or biological constraints bear traces of cultural attitudes about reproduction and motherhood. Ovaries become objects for removal through a complex chain of scientific statements enabled by the singular focus of cancer researchers on risk reduction and through the signification of ovaries as mere reproductive organs—all of which is further complicated by the racial framework in which such knowledge claims emerge and circulate.

Ovarian Cancer and Heredity

Ovarian cancer stands out among the reproductive cancers affecting women. According to the American Cancer Society, in 2004 approximately 26,000 women were diagnosed with ovarian cancer, and 16,000 did not survive.[4] The National Cancer Institute predicts that in 2011, 22,000 women will have been diagnosed and 15,500 will have died.[5] Estimates for 2012 are basically the same.[6] Indeed, ovarian cancer causes more deaths than any other cancer of the female reproductive system.[7] As is the case with many other diseases, the statistics also reveal disparities. Black women, for instance, are less likely to get ovarian cancer than white women, but they are more likely to die from the disease (a pattern similar to that of breast cancer). Moreover, trends suggest that while incidence and mortality is declining overall for white women, for black women both have remained steady.[8]

The consensus in the literature is that early detection will not reduce mortality rates to an appreciable degree. Although technologies such as transvaginal sonography and the cancer antigen blood test known as the CA-125 screen are available, the evidence is mixed as to whether they can reliably and consistently detect the disease at an early stage and improve survival (ovarian cancer is very deadly once it progresses beyond stage I—when it is still quite localized). As a result, these detection procedures are not routinely performed, unlike mammography for the early detection of breast cancer and pap smears for the early detection of cervical cancer. Most ovarian cancers, then, are diagnosed at stage II or later, the result being an extremely high death rate. Identifying women at higher-than-average risk so that they can undergo detection procedures has, understandably, been a goal of cancer researchers and at-risk women.

Well before the first BRCA gene was mapped, interest in a genetic predisposition for ovarian cancer was sparked by knowledge of the so-called hereditary cancer syndromes. These syndromes explained why in some families, many women—especially young women—were diagnosed with breast and/or ovarian cancer. A study based on the Gilda Radner Familial Ovarian Cancer Registry from 1981 to 1991 established that ovarian cancer risk could indeed be understood as a genetic predisposition for some families.[9] In those cases where two or more first- and second-degree relatives were diagnosed with ovarian cancer, risk was apparently linked to an autosomal dominant trait. The researchers studying families in the Radner Registry found that the vast majority of such cases were mother-daughter and sister-sister pairs.[10] The risk for women with a family history that included mother-daughter cases was estimated to be about 41 percent but could be as high as 50 percent.[11] Participants in a workshop sponsored by the National Cancer Institute and the National Center for Human Genome Research (now called the National Human Genome Research Institute) concluded that 1 percent of women with ovarian cancer have a family history consistent with an autosomal dominant mode of transmission, although an additional 7 percent of women with ovarian cancer and a family history of at least one affected first- or second-degree relative are thought to be at moderate risk for the disease.[12] "Ultimately," concluded the study based on the Radner Registry, "early detection of the individual at risk for ovarian cancer will depend on finding the abnormal gene."[13]

As noted in chapter 2, Steven Narod and colleagues reported that ovarian cancer risk was linked to mutations of BRCA1.[14] Several large studies of ovarian cancer have demonstrated that mutations of BRCA1 and BRCA2 can elevate a woman's lifetime risk for the disease.[15] Other studies of women with ovarian cancer—with or without a family history of the disease—have revealed that as many as 10 percent test positive for mutations, thus revising significantly the prior estimate of 1 percent.[16]

Women with BRCA mutations face a 10–60 percent chance of developing ovarian cancer, depending on which mutation they inherit, with BRCA1 mutations generally conferring a higher risk and accounting for an earlier age of onset. Treatment recommendations for high-risk women have often included oophorectomy,[17] largely because ovarian cancer is such a deadly disease. In 1995 the National Institutes of Health Consensus Development Panel on Ovarian Cancer concluded:

> The probability of a hereditary ovarian cancer syndrome in a family pedigree increases with the number of affected relatives, with the number of affected generations, and with young age of onset of disease. There, prophylactic oophorectomy should be considered in these settings with careful weighing of the risks and benefits. The risk of ovarian cancer in women from families with hereditary ovarian cancer syndromes . . . is sufficiently high to recommend prophylactic oophorectomy in these women at 35 years of age or after childbearing is completed.[18]

In 1999 officials from the New York State Department of Health wrote that "if a patient at risk for ovarian cancer is having abdominal/pelvic surgery, a prophylactic oophorectomy should be performed."[19]

Not all recommendations for prophylactic oophorectomy have been so strongly worded, largely because scientific evidence demonstrating the efficacy of oophorectomy was limited prior to 2002. Particularly problematic was the fact that no prospective studies comparing women electing surgery with women electing surveillance had been carried out, and in the retrospective studies that did show a benefit from the surgery, the subjects did not have BRCA mutations. A Consensus Statement by the Cancer Genetics Studies Consortium organized by the US National Human Genome Research Institute did not include an explicit recommendation for oophorectomy, writing that "observational data have so far failed to demonstrate statistically significant evidence for risk reduction."[20] Much of the concern over oophorectomy was the result of researchers' observations that up to 10 percent of women undergoing oophorectomy were diagnosed with ovarian-type cancers years later.[21] These women were diagnosed with peritoneal carcinoma, a disease indistinguishable from ovarian cancer.[22]

Qualified language concerning prophylactic oophorectomy was quickly replaced with enthusiastic assessments after the publication of two studies demonstrating that ovarian cancer risk can be lowered after oophorectomy for BRCA carriers. In one study, by Noah Kauff and coauthors, after a mean follow-up of two years, one out of ninety-eight women who had oophorectomy developed cancer (peritoneal), compared to five out of seventy-two women who elected to undergo surveillance only (four had ovarian cancers, one peritoneal cancer). In the other study, by Timothy Rebbeck and coauthors, 2 out of 259 women undergoing oophorectomy were diagnosed with peritoneal cancer during the approximately eight years of follow-up, compared to 58 out of 292 women who opted for surveillance only and were later diagnosed with ovarian cancer. For the Kauff group, the results provided "strong support for including discussion of risk-reducing salpingo-oophorectomy as part of a preventive-oncology strategy for women with a BRCA1 or BRCA2 mutation."[23] Rebbeck and coauthors wrote: "On the basis of the results of our study, we advocate prophylactic oophorectomy to reduce the risk of ovarian and breast cancer in women with BRCA1 or BRCA2 mutations."[24]

These 2002 oophorectomy studies are widely cited in the cancer genetics literature as scientific evidence that the procedure is, on the whole, beneficial for women with BRCA mutations. Writing in the *Journal of Clinical Oncology* in 2003, Douglas Levine and colleagues claimed that "risk-reducing salpingo-oophorectomy is currently recommended for

BRCA mutation carriers after completion of childbearing or at age 35 years."[25] A 2004 literature review published in the journal *Gynecologic Oncology* concluded: "Prophylactic surgery conveys the greatest risk reduction and is a procedure that needs to be offered to all BRCA1 and 2 germline carriers at this time."[26] And a 2004 article in the *Journal of Clinical Oncology* concluded that, given the available data regarding risk reduction from oophorectomy, the question for women with BRCA mutations is no longer "whether" to have the procedure but "when."[27] For cancer researchers and practitioners, these studies provide clinical evidence that surgery can reasonably be expected to manage risk that only BRCA tests can quantify with any certainty.

Oophorectomy in Historical Context

A BRCA test may result in mere statistical probabilities, yet the surgical interventions that follow are thinkable not only because of the outcome of risk assessment, but also because of assumptions about the physiological role of ovaries in health and disease. The fact that a BRCA test can result in the removal of a woman's organs necessarily raises the question of gender in science and medicine. Conceptual frameworks for understanding ovarian function inevitably rely on social values related to reproduction, motherhood, and the social order. As we will see, the field of gynecology has performed important ideological work over time, linking oophorectomy, gender, and hereditarian thinking far earlier than their recent convergence in BRCA research.

In part this is due to what historians have documented as the scientific and symbolic importance of the ovary in the Western, industrialized world. Ovaries are material artifacts that encompass and condense historically varying and multilayered ideas about gender, race, and reproduction. These ideas shape gynecological practice; more important, when circulated back into public discourse, they shape cultural beliefs about the naturalness of sex and gender. Moreover, as a surgical procedure, oophorectomy raises important questions about the relationship between genomics and clinical medicine—in this case, the ability of oophorectomy to satisfy the needs of multiple interests simultaneously.

The symbolic and scientific significance of the ovary has a long history. Ovary removal, whether for treatment or prophylactic purposes,[28] has played a prominent role in the development of surgical practice and, more specifically, gynecology in both the United States and the United

Kingdom. It has also played an important part in the emergence of medicine as a key site for the surveillance and disciplinary control of women's bodies. To properly situate contemporary oophorectomy historically requires an examination of the many justifications for ovary removal over time—justifications that reflect social values concerning gender, reproduction, and motherhood. Ornella Moscucci has noted that gynecology is a "specialism which is underpinned by a historically contingent notion of woman."[29] In turn, gender roles have obtained social legitimacy by way of elite medical discourse.[30] Specifically, the ovary was, in the nineteenth century, key to the reigning model of femininity.[31]

The first successful ovariectomy, attributed by many historians to Ephraim McDowell, a physician practicing in Danville, Kentucky, has also been described as the "first successful elective exploratory laparotomy in the world."[32] In 1809 the unidentified female patient endured what was acknowledged by McDowell to be an experimental surgery without the benefit of anesthesia or antisepsis (the former was not available, the latter not considered important) to remove a large ovarian cyst. Subsequently, throughout the early 1800s, surgeons pursued their interest in ovariectomy, largely in the effort to advance their craft and establish gynecology as a legitimate field of medicine and to treat ovarian disease that they argued was grave enough to warrant life-threatening surgery. "During [oophorectomy's] heyday doctors boasted that they had removed from 1,500 to 2,100 ovaries apiece."[33]

Critics, however, were quite forceful in their attacks, calling these surgeons "belly-rippers," responsible for the death of women with medical conditions that were not necessarily fatal.[34] "By the mid-1800s," writes Moscucci, "ovariotomy was in dispute."[35] Nevertheless, the procedure continued to be performed, as it was both lucrative and central to the professionalization of gynecology—at the time a young specialty. Specifically, ovariotomy helped gynecology emerge as a reputable and indispensable area of medicine that could compete with the already established field of obstetrics.[36] Mortality rates from ovariotomy gradually declined, in part due to the development of effective antisepsis. Yet the most significant development was the introduction of anesthesia,[37] a development that undermined critics' arguments that the surgery was, in effect, a form of inhumane experimentation on women.

Practitioners at this time also drew connections between women's reproductive organs and their "nature" in order to secure for gynecology a role in the rather lofty debate about gender and society. Referring to the

British gynecologist William Blair-Bell, Chandak Sengoopta writes that the belief that the female was reducible to reproductive biology "was not simply a patriarchal conviction but was also crucial to Bell's lifelong effort to elevate the status of gynaecology. If woman was wholly and exclusively governed by her biology, then that would be the best justification for the existence and the advancement of the gynaecological profession, which, according to many elite generalists, was an example of the dangerous proliferation of medical specialties."[38]

The historical significance of ovariotomy, then, had less to do with the successful treatment of ovarian disease than with the development of surgical procedure, the latter being crucially important to the field of gynecology. "By 1870," writes Moscucci, "ovariotomy had become established as an accepted procedure and was hailed as the triumph of gynaecology and the beginning of abdominal surgery."[39] Gynecology would eventually be a successful case of medical specialization, a field dependent more on the heroics of surgery for its growth than with the rigors of research.[40] The physician John Studd observes that "each generation of surgeons (gynecologists in this case) has had a fashionable operation which taught the trainee surgeon how to open and close abdomens although in retrospect it can be seen that the operations were either useless or superseded by a better method."[41] Eventually, "ovariotomy became the measure of a surgeon's ability at a time when most advances in abdominal surgery were performed in women with gynecological disorders. These disorders were both real and imaginary."[42]

The "imaginary" disorders to which Studd refers, in conjunction with the use of anaesthesia, largely accounted for the performance of ovariotomy for nonmedical reasons. Surgeons would "remove ovaries that were 'prolapsed' [displaced] but otherwise healthy" and for "nebulous conditions which are not nowadays regarded as pathological states, for example 'ovarian cyrrhosis.'"[43] Eventually, the term "normal ovariotomy" was coined, denoting the removal of ovaries to treat peripheral diseases.[44] The surgery would become very popular in the United States and in Europe for conditions such as "menstrual madness, oophoromania,[45] hysterical vomiting, epilepsy, dysmenorrhea,[46] and those great Victorian disorders of nymphomania and masturbation. Leeches had been applied to the lower abdomen, vulva and anus for these symptoms for decades but from 1880 this treatment had given way to the removal of ovaries, in order to prevent insanity and moral decline."[47] In some cases, all instances of mental illness were linked to ovaries: "The results [of normal ovariotomy] were apparently so successful that it was soon performed for 'all cases of lunacy' and young surgeons would be given an annex of a psychiatric hospital where they would remove ovaries from the inmates."[48]

The emergence of the so-called normal ovariotomy is but one manifestation of the general belief that reproductive organs were central not only to a woman's identity, but also to her physical and mental health, broadly defined. Importantly, the health of white, middle-class women was viewed as crucial to the security and stability of the nation. Indeed, early eugenicist thinking pervaded gynecological discourse, especially with regard to the connection between ovarian disease and mental illness: "From one perspective, the castration of women starting in the 1870s (performed overwhelmingly on noninstitutionalized outpatients, and only later—in the 1890s—on inmates of mental institutions) was part of the general anxiety about the racial future of white America."[49] Articulating the social importance of gynecology, Lawson Tait, one of "normal" ovariotomy's namesakes, observed: "To the naked eye nothing could look more uninteresting and unimportant than the human ovary; and yet upon it the whole affairs of the world depend."[50]

What exactly threatened the ovary and, in turn, the social order in the United States in particular? For public intellectuals, cultural critics, and physicians alike, white, middle-class women's push for education, entrance into the public sphere (to initiate and participate in reform campaigns, most notably for temperance and the franchise and against slavery), and fight for reproductive autonomy called into question popularly held beliefs about gender.[51] Contested ideas about women, motherhood, and femininity further threatened national identity—indeed, the nation itself. An educated woman, they argued, was less likely to reproduce, and women who were involved in the political realm were likely to neglect their expected duties in the private sphere, thereby jeopardizing the next generation of citizens and leaders. Especially during the latter part of the nineteenth century, worries over the American woman's physical fitness stirred fears of race suicide: "The young women of the urban middle and upper classes seemed in particular less vigorous, more nervous than either their own grandmothers or European contemporaries."[52] The putative "deterioration" of women "called gynecology into existence," according to the physician Augustus Kinsley Gardner.[53]

Scientific discourse further served a critical disciplinary function: "Since at least the time of Hippocrates and Aristotle, the roles assigned women have attracted an elaborate body of medical and biological justification.

This was especially true in the nineteenth century as the intellectual and emotional centrality of science increased steadily. Would-be scientific arguments were used in the rationalization and legitimization of almost every aspect of Victorian life and with particular vehemence in those areas in which social change implied stress in existing social arrangements."[54] Medical dogma at the time held that brains and reproductive organs could not thrive simultaneously, thus providing a scientific justification for the denial of women's civil and human rights. Specifically, it held that a woman who pursued an education was likely to suffer from uterine and/or ovarian disease and risked becoming an unfit mother. Moreover, a woman who did not procreate was destined to suffer diseases of the reproductive organs stemming from nonuse: "The alleged incidence and seriousness of ovarian disease legitimated the view that sex and reproduction dominated women's body and mind."[55] Reproductive organs, used or unused, posed lifelong threats to a woman's health: thus, all women were at risk, and all were subject to the disciplinary discourses of medicine. The historians Carroll Smith-Rosenberg and Charles Rosenberg capture the double bind of medical theories of the reproductive system: "Physicians saw woman as the product and prisoner of her reproductive system. It was the ineluctable basis of her social role and behavioral characteristics, the cause of her most common ailments; woman's uterus and ovaries controlled her body and behavior from puberty through menopause."[56]

The cultural imperative to reduce women to their biology elevated the status of the ovary in medical theories about gender and health. For the ovary enabled a more comprehensive explanation of women's health and degeneracy than the uterus, because the former's hormonal function was more generalizable. Whereas diseases of the uterus were localized—limited to the uterus itself—the ovary was physically linked to the brain and other bodily organs due to its regulation of hormones.[57] The "reflex irritation" theory held that the reproductive organs were inextricably linked to the nervous system—an "irritation" to one (for instance, education) would inevitably have an impact on the other (the nervous system or psyche).[58] These theoretical and clinical developments were in turn critical to the introduction of oophorectomy as medical practice and for advancing the eugenicist agenda by effectively sterilizing genetically unfit women. The paradox, of course, was that oophorectomy robbed too many women of the very identity and reproductive capacity necessary for nascent eugenicist projects. Thus, critics of the procedure—especially in the United Kingdom[59]—successfully deployed arguments about women, motherhood, and nation to reduce the number of surgeries performed. Male physicians and women activists benefited strategically from the symbolic importance of ovaries, arguing that ovariotomy stripped women of their identity (the charge of "desexing" women was common), an identity largely rooted in motherhood and its connection with nationalism and the imperialistic imaginary:[60] "Flagrant abuse of the procedure, its hazards and questionable results, but above all the gradual realization that ovaries are precious organs [William Goodell presiding (over) the 3rd annual meeting of the American Gynecological Society in 1875 thought fit to remind his audience that 'woman is so called because she bears a womb' and 'physiologically she is a woman because she owns two ovaries'] caused the pendulum to swing and initiated the downfall of Battey's operation."[61]

The historical narrative I have thus far constructed shows how oophorectomy has served the institutional interests of surgeons (specifically gynecologists) since the early part of the nineteenth century. The procedure also figured centrally in debates about women, their reproductive organs, and the social order at large. When ovariectomy for explicitly cultural reasons fell out of favor, ovarian cancer would present gynecologists with the opportunity to fight a deadly disease requiring nothing less than radical intervention. A positivist perspective, informed by a logic of progression, would hold that this transition to cancer prevention is simply the expected outcome as gynecology passed over into a more enlightened period of its history. A social, materialist history of gynecological surgery asks instead what this new reconfiguration of objects, patients, practices, and social relations actually means in terms of particular interests served.

Oophorectomy's New Legitimacy Crisis

By the turn of the century in both the United States and many European nations, "only diseased ovaries tended to be removed."[62] The meaning of "diseased" was open to interpretation, however, and ovariectomy was still performed in order to remove benign growths. Moreover, what some historians have described as the "discovery" of ovarian cancer as a public health threat lent support for the prophylactic removal of ovaries even by those who had previously supported ovarian conservation. The insidiousness of ovarian cancer (frequently characterized as a form of so-called creeping death), more than its actual incidence, accounted for the revival of oophorectomy. As one gynecologist wrote in 1942, "the involuting[63] ovaries have fulfilled their reproductive and endocrine function. They are no longer an important part of the economy but vestigial structures which carry a special tendency to cancer."[64]

As a result of this thinking, oophorectomy is now routinely performed during hysterectomy or other abdominal surgery for noncancerous, "benign" pathological conditions.[65] In the United States alone, approximately 300,000 prophylactic oophorectomies are performed each year, with the reduction of cancer risk being the primary reason for the procedure.[66] In the United Kingdom in 2003, 19,000 women had elective bilateral oophorectomies.[67] The percentage of hysterectomies during which bilateral salpingo-oophorectomy is performed more than doubled from 1965 (25 percent) to 1999 (55 percent).[68] And "data from the Centers for Disease Control and Prevention show that 38 percent of women have concurrent [with hysterectomy] oophorectomy between ages 18 and 44 and 78 percent between ages 45 and 64."[69] An overview of the practice in the United Kingdom concluded: "Prophylactic removal of the colon, the breast and the appendix does occur, but oophorectomy at the time of hysterectomy represents the highest population-based rate of removal of any healthy organ for prophylactic reasons."[70]

Medical education is an important source of the popularity of oophorectomy among physicians. As David Olive notes, "the predominant teaching is that prophylactic oophorectomy in the low-risk patient should be avoided under the age of 40, should be routinely performed over age 55, and should be considered and discussed in the interval between."[71] Further evidence comes from a survey of members of the Royal College of Obstetricians and Gynaecologists in the United Kingdom, which revealed that "85 percent [of respondents] would consider the routine removal of both ovaries in a postmenopausal woman at the time of abdominal hysterectomy. Similar findings have been observed in surveys of physicians in the United States, Ireland, and Italy."[72] These respondents believe that abdominal surgery presents a unique opportunity for cancer prevention. Indeed, one set of guidelines cited a report "that elective routine removal of ovaries after age 40 would eliminate approximately 12–14 percent of all ovarian cancers."[73]

The rationale (or "dogma," according to some critics)[74] is this: if a woman is undergoing abdominal surgery—whether or not she has been diagnosed with benign ovarian disease—oophorectomy ought to be recommended to her if she no longer wishes to be pregnant and/or is postmenopausal or perimenopausal (oophorectomy for premenopausal women at average risk for ovarian cancer is more controversial and is discussed below). For example, the authors of a 1997 review wrote that "prophylactic oophorectomy during benign pelvic/abdominal surgery is recommended for women who have a significant family history of ovarian cancer and have completed child bearing or for women aged >40 years."[75] Such a recommendation accommodates the desire of some women to have children; it further rests on the assumption that postmenopausal ovaries are no longer an important component of a woman's endocrine system. In short, after menopause, ovaries are superfluous and can do only harm: "If the woman is already menopausal [and undergoing pelvic surgery for another condition], it [is] well worth counseling her of the entity of ovarian cancer, as many are not aware of the condition. There is *no good reason to retain menopausal ovaries*, as the risk of developing ovarian cancer increases with age" (emphasis added).[76]

Although no physician would recommend that a woman at average risk for ovarian cancer undergo surgery for the sole purpose of removing her ovaries,[77] the practice of removing them if the surgical occasion presents itself is widely accepted. This belief is underscored by concern that, often, the condition or conditions necessitating surgery in the first place are part of a generalized disorder, sometimes called "residual ovarian syndrome,"[78] that will probably require additional major surgeries in the future, thereby exposing women to unnecessary risks (including those posed by anesthesia), significant recovery time, and, perhaps, decreased quality of life.[79] "Since the 1950s," observed two cancer researchers, "the issue of prophylactic oophorectomy has been plaguing the surgeon. Studies have confirmed that 4.5–14.1 percent of women develop ovarian cancer after hysterectomy for non-ovarian conditions. Similarly, data from the American College of Surgeons demonstrated that among 12,316 patients with ovarian cancer, 18.2 percent had previous hysterectomies with conservation of one or both ovaries and 57.4 percent of these women were over the age of 40."[80]

In 2005, however, the rationale for routine oophorectomy at the time of abdominal surgery was called into question by the publication of a decision model (a method of cost-benefit analysis) showing that prophylactic oophorectomy heightens risks of death from heart disease and osteoporosis, risks significant enough that they probably outweigh the benefit of lowered risk for ovarian cancer.[81] The researchers—who were from the University of California, Los Angeles; the University of Southern California; and the University of Auckland—claimed that the study was necessary because although the procedure was performed on a great many women, little information existed regarding its relative advantages and disadvantages. A previous decision model by Katrina Armstrong and coauthors had shown some benefit, but it pertained to BRCA carriers only, not to women in the general population at average risk (the exceptional status of BRCA carriers is discussed in the next section).[82]

The 2005 decision model by William Parker and coauthors sought a more nuanced assessment than previous models by analyzing key variables for the first time. It assumed, for instance, that—based on published research—ovarian cancer risk is reduced significantly by hysterectomy alone. Second, it assumed that ovaries perform important endocrine functions; for example, even after menopause, ovaries continue to produce significant amounts of testosterone and androstenedione, androgens that are then converted to estrogen.[83] This decision model further assumed that women are not likely to undergo hormone-replacement therapy (HRT) to compensate for estrogen loss resulting from oophorectomy: not only does HRT have many undesirable side effects, but results of the much-publicized Women's Health Initiative cast doubt on the ability of HRT to reduce the risks of cardiovascular disease. According to the research team's conclusion, the excess death predicted from prophylactic oophorectomy is considerable:

> For a hypothetical cohort of 10,000 women undergoing hysterectomy who chose oophorectomy between the ages of 50 and 54 without estrogen therapy, our analyses predict that, by the time they reach age 90, 838 more women will have died from CHD [coronary heart disease] than in a similar cohort of women who chose ovarian conservation; 158 more will have died from hip fracture; 47 fewer women will have died from ovarian cancer. In the base case analyses, oophorectomy in women ages 50–54 leads to an overall excess mortality of 858 per 10,000 women subjected to surgery.[84]

The authors concluded: "The model shows that women younger than 65 years of age clearly benefit from ovarian conservation, and at no age is there clear benefit from oophorectomy."[85]

Although previous research had documented the adverse health effects of oophorectomy,[86] the publication of the study by Parker's group sparked considerable controversy. In a 2006 issue of the journal *Climacteric*, a University of Cape Town researcher challenged the decision model's operating assumptions (for example, that few women would have HRT) and concluded that "the decision analysis reviewed here is not informative, and it offers no guidance to gynecologists in deciding whether or not to remove the ovaries when performing hysterectomy."[87] Davy and Oehler claimed it was premature to dismiss post-oophorectomy interventions that could help reduce long-term mortality, such as HRT, statin therapy (for reducing cholesterol), and bone density correctives, warning that in comparison "there is no prevention for ovarian cancer but oophorectomy."[88] They continued: "It is unlikely that many surgeons are going to change their management based on the current controversy."[89] John Studd, the physician openly critical of late-nineteenth-century routine ovariectomy, took a somewhat different approach and, in addition to championing modern medicine's ability to prevent or reduce the morbid effects of prophylactic oophorecotmy, characterized advocates of the procedure as courageously sticking to principle in the face of "fashionable" and "politically correct" criticism (he claimed in 1989 that oophorectomy should be performed for all women over the age of forty when having abdominal hysterectomy).[90] Studd's most pointed attack was directed at the assumption that women do not comply with recommended HRT schedules. It is the responsibility of the surgeon, he argued, to deal aggressively and vigilantly with women who resist the therapy. In response to the preponderance of evidence that side effects of HRT accounted for women's reasonable resistance to it, Studd wrote: "Even if there are side-effects to replacement therapy, it would be *hard to believe* that estrogens are more dangerous than conserved ovaries with their malignant potential as well as their endogenous production."[91] He concluded by calling for randomized trials as a way to definitively determine the benefits of prophylactic oophorectomy for women at average risk, even though he surely knew no study of the kind would ever pass ethical muster.[92]

Still, Parker and his colleagues did have their supporters, suggesting that the model would, at the very least, motivate the medical community to reexamine the practice of routine oophorectomy. One physician wrote: "The decisions by patients and their doctors can now be based upon better facts rather than just 'what is done.'"[93] David Olive echoed this sentiment in an editorial in *Obstetrics and Gynecology*, adding that clinical practice would probably be altered (a change that might be "time-consuming and annoying to some"), an important shift because "even if no coronary heart disease increase is seen with oophorectomy, there remains no demonstrable advantage to the procedure in terms of longevity!"[94]

Moreover, Parker and coauthors responded to their critics, writing: "We hope that women and their physicians will discuss the long-term implications of oophorectomy and that many physicians will recommend ovarian conservation for women at average risk of developing ovarian cancer."[95] Significantly, they reiterated their doubt that HRT was unquestionably the solution to long-term mortality, adding that their decision model predicted unacceptable mortality even if 100 percent of women engaged in estrogen therapy until the age of eighty. Lower rates of use (a more realistic assumption) "would proportionately increase mortality and, therefore, oophorectomy is a disservice for the vast majority of these women."[96] (Indeed, HRT rates have declined significantly with ongoing publication of results from the Women's Health Initiative study showing that HRT probably increases, not decreases, the risk of heart disease and that it significantly increases breast cancer risk).[97] The Parker team asked their critics the provocative question (highly relevant, given the history of this surgery): "What ever happened to 'evidence' and the principle 'first, do no harm'?"[98]

A retrospective cohort study published in 2006 by Walter Rocca and coauthors provided additional evidence challenging the advisability of current medical thinking about oophorectomy. Specifically, the authors concluded that women under the age of forty-five who undergo oophorectomy and refuse HRT face a significant risk of death.[99] Using methodology similar to that employed by the Parker team, the Rocca group did not examine women with BRCA mutations or with significant family histories of the disease, only women at average risk. Comparing women who had oophorectomy with those who did not, they found that young women undergoing the procedure experienced an increased risk of "oestrogen-related cancer," such as breast cancer.[100] These women also faced the chance that they might eventually die from oophorectomy-related cardiovascular disease, osteoporosis, and neurological diseases such as dementia and Parkinson's disease.[101] The authors concluded: "For most women with prophylactic bilateral oophorectomy before age 50 years who died of a non-cancer cause, ovarian conservation could have provided a survival advantage."[102] Despite the more conservative language of the Rocca group's report,[103] several physicians predicted that the study's results would have an impact on clinical practice.[104] In 2008 ACOG (also known as the American Congress of Obstetricians and Gynecologists) updated its guidelines on elective salpingo-oophorectomy, urging "strong consideration" for "retaining normal ovaries in premenopausal women who are not at increased genetic risk for ovarian cancer."[105] According to

Esther Eisenberg of the North American Menopause Society, "for most women, the scales are now tipping towards ovarian conservation."[106]

Emily Martin has written that it is easier to see the connections between science and society from the privileged vantage point of the historian.[107] Nevertheless, these connections remain in the contemporary moment. To be sure, the most recent oophorectomy controversy forces us to question positivist histories of science and medicine and the logic of progress on which they depend. The routinization of oophorectomy for cancer prevention is no less a sociopolitical phenomenon than was routine oophorectomy in the nineteenth century. What the entire history of oophorectomy tells us is that changes in clinical practice are inextricably linked to societal attitudes about the body at risk as well as to the conceptual and professional needs of researchers and physicians. In the early part of the nineteenth century, the availability of women's bodies (largely the result of prevailing ideas about reproductive organs, motherhood, and gendered frameworks of disease assessment), coupled with the need to advance surgical techniques and solidify the relevance and sophistication of gynecology, made oophorectomy a popular—and abused—procedure.

Today the state of scientific knowledge about ovaries plays a significant role in making their removal thinkable largely because the ovary is metonymically reduced to its reproductive role. As the authors of a research review concluded, "the normal physiology of the post-menopausal ovary has not been extensively studied, and it is possible that removal of the post-menopausal ovaries may have greater consequences than has previously been believed. It is clear that the effects of bilateral oophorectomy in pre- and post-menopausal women at the time of hysterectomy are a fertile ground for clinically significant research."[108] In other words, the scientific community is only now beginning to take seriously the role of ovaries in a general economy of health, suggesting that these organs have remained both the biological and the symbolic seat of femininity and motherhood longer than has been acknowledged. Nevertheless, oophorectomy will continue to capture the attention and interest of gynecologists and cancer specialists as it has a new constituency: the BRCA mutation carrier.

Constitutive Discourse and the Medical Subject

As the revised ACOG guidelines illustrate, the gynecology community has recognized the need to rethink oophorectomy. But ACOG—as well as both the Parker and Rocca research teams—emphasizes that treatment for

BRCA carriers should not be affected by recent developments.[109] Although women at average risk may now be counseled differently about ovarian conservation, oophorectomy is unequivocally advocated for women bearing BRCA mutations. For this reason, BRCA testing has proved institutionally significant. Oophorectomy provides much-needed legitimacy for the field of cancer susceptibility testing insofar as it provides an evidence-based method of acting on inherited risk and helps temper the potentially explosive criticism of gynecology that recent research on oophorectomy for non-BRCA carriers has elicited.[110] The questions we are confronted with now are, why do BRCA carriers constitute an exception to the debate about prophylactic ovarian removal, and how is this exceptional status made possible?

As I discussed above, mapping the BRCA genes was an important medical development, as women from cancer syndrome families in which ovarian cancer was common had long been advised to consider prophylactic oophorectomy. These women found themselves contemplating major surgery even though they could not be sure they had inherited a mutation; moreover, they had to make their decision with little evidence that oophorectomy reduced risk enough to justify the operation. In the mid-1990s, the new ability to screen for the BRCA mutation made it all the more imperative to document the effectiveness of existing prevention measures. As I discussed earlier in the chapter, in 2002 two studies provided such documentation. Since then, a clear consensus around oophorectomy has emerged.

The received view holds that these 2002 studies merely provided, for the first time, concrete evidence of the efficacy of prophylactic oophorectomy in decreasing risk for ovarian cancer, particularly for women with BRCA mutations. I want to question this interpretation, especially insofar as it rests on the common-sense understanding that the BRCA test produces more actionable knowledge about risk than either an analysis of family pedigree analysis or predictions from a cancer registry can. It is different information, to be sure, arising from the material analysis of a woman's bodily tissues. Yet its conceptual and epistemological significance lies not in how it produces accurate assessments of actually existing risk. That significance rests, rather, on the way the BRCA test substitutes the mark of ancestry for the materiality of lived experience, removing from the clinician's gaze a range of mitigating and aggravating variables that would otherwise influence what counts as actionable risk.

The studies we now have about the effectiveness of prophylactic surgery in reducing risk were carried out well after ovarian cancer was classified as a heritable trait. In fact, their very purpose was to address the concerns of cancer physicians that genetic testing would fuel interest in the procedure. One physician wrote: "The identification of the BRCA genes and the availability of genetic testing for BRCA mutations have underscored the need for improvements in early detection and prevention of breast and ovarian cancer."[111] And since the studies by Kauff and colleagues and Rebbeck and colleagues showed that BRCA mutation carriers in particular would benefit, calls for more genetic testing have followed.[112]

Geneticists distinguish BRCA tests from analyses of family history in a number of important ways. Unlike family pedigrees, they say, BRCA tests bring "the promise of reducing uncertainty surrounding [BRCA carriers'] risk status, thus enabling them to make more informed decisions about whether to undergo prophylactic surgery" (emphasis added).[113] According to the authors of a review essay, recent insight into the hereditary nature of some cancers and the advent of genetic testing have made risk assessment increasingly accurate."[114] Family history, once an important factor in assessing a woman's risk, has increasingly become regarded as inaccurate and inferior. Writing in the *Journal of Clinical Oncology*, researchers from cancer centers in the United States and Canada concluded that "the strongest risk factor for ovarian cancer is the presence of an inherited mutation in 1 of the 2 ovarian cancer susceptibility genes, BRCA1 or BRCA2."[115] As one author put it, family history alone is a "poor predictor."[116]

Yet the risk associated with BRCA mutations, much like the concept of risk more generally, is not objective fact, but a social construct.[117] Both family pedigree data and BRCA tests produce statistical probability statements—the *telos* of both types of analysis is to establish risk, not diagnose disease. They tell a woman, based on clinical evidence, whether she is at significant risk and, thus, whether she should consider radical therapies. Before the BRCA test, risk was based on data from registries as well as familial and environmental history. In the wake of the test's commercialization, risk is now calculated based on the structural features (and presumed function) of the mutation that the woman has inherited. Clinical data are limited to the study of only some of the thousands of known mutations. Risk assessment, then, is the determination that the chemical makeup of the mutation does or does not suggest that it is "deleterious,"[118] not necessarily whether clinical data do or do not demonstrate the approximate risk it confers.

Regardless of the method used, both types of risk assessment presume risk to be variably penetrant. Before genetic testing, women with a strong family history of ovarian cancer were told their risk was as high as 50 percent. BRCA-related risk can be as low as 10 percent and as high as 60 percent depending on the mutation, where the mutation is located along the gene,[119] and a whole host of environmental factors.[120] Despite this variable penetrance, BRCA tests nevertheless occupy a privileged diagnostic status, and this in turn has shifted the evidentiary grounds on which decisions about prophylactic surgery are made. The presence of a mutation has become the indisputable basis for action and thus operates as a distinct epistemological category in the discourse of cancer genomics. Today, women with BRCA mutations are unqualified candidates for prophylactic surgery, whereas before BRCA tests, caveats concerning the advisability of surgery were part and parcel of recommendations for women with a documented family history of ovarian cancer. Despite the fact that both types of risk assessment produce statistical probability statements, inherited ovarian cancer discourse betrays a theoretical transformation that temporally and conceptually collapses the movement from test and surgery. Risk, one might say, *is* the disease.

Our fate, then, *is* in our genes, as Francis Collins once famously proclaimed. Indeed, the theoretical and rhetorical conditions whereby treatment recommendations for women *at risk* for ovarian cancer and women *with* ovarian cancer are the same can be traced to the long-standing status of heredity in our culture. More than this, though, is the privilege that a particular type of vision holds in medicine. Implicitly, genomics benefits from visual culture insofar as the DNA base pairs of the genetic test allow the diagnostician to "see" the very thing that increases a woman's risk for cancer. [121] Science studies scholars have documented the importance of visual imagery in medicine,[122] an instantiation of the importance of vision more generally in Anglo-American culture.[123] BRCA tests, we are told, reveal "actual risk"[124] of "genetically defined"[125] women. Yet it is not at all clear what exactly distinguishes what is "seen" in one's family history from what is "seen" in one's genome, other than a shift in the field of vision. What the BRCA test performs is the conceptual removal of a woman from a particular context or environment and, with that, a consideration of embodied life in all of its complexity.

Acting on genetic risk is further justified in biomedical discourse on the grounds that no other method of surveillance can protect a woman with a BRCA mutation. Prior to the development of oncogenomics, for the vast majority of members of cancer syndrome families, risk justified intense surveillance, not prophylactic surgery. In contrast, the BRCA test reveals putatively real risk and, by finding risk, replaces traditional screening methods such as sonography and CA-125 serum testing, which can detect only actual cancer. BRCA tests have become, in the words of cancer researchers, substitutes for screening: they are a method of true prevention.[126] Moreover, cancer surgeons have reported finding ovarian cancer tumors during oophorectomy, bestowing on the procedure an entirely different, unexpected purpose: early detection. As such, it has provided material evidence for the cancer surgeon that genetic risk is never merely risk.

The elevation of BRCA tests as a form of primary (and thus superior) surveillance is inextricably linked to the routine dismissal of the alternatives. It is not uncommon for researchers to discount the efficacy of secondary screening despite conflicting evidence regarding the effectiveness of sonography, especially for women with a family history of ovarian cancer.[127] At best, researchers characterize the extant data as mixed; at worst, they claim that "*no* screening for ovarian cancer has yet been proved effective for women in *any* risk category" (emphasis added).[128] In 2006, shortly after the publication of the Parker decision model, the endorsement of oophorectomy became even more emphatic: "Ovarian cancer is, in most cases, a lethal disease as it is virtually impossible to diagnose at an early stage, and almost impossible to treat successfully when detected at an advanced stage. Thus, *any means* we have of reducing the incidence of the disease should be embraced" (emphasis added).[129]

Yet as early as 1991, some researchers were reasonably optimistic about the benefits of another kind of screening for women at high risk:

> TVS [transvaginal ultrasonography] is particularly effective as a screening method in women whose primary or secondary relatives have documented ovarian cancer. By limiting screenees to those with a positive family history of ovarian cancer, the authors found that the prevalence of ovarian cancer in the screened population increased 10-fold. As a result, the positive predictive value of TVS increased proportionately. Three primary ovarian cancers were detected in 776 asymptomatic women. All three women had Stage I disease and all are now alive and well.[130]

Conceivably, the use of TVS in the above study presumes that hereditary ovarian cancer is a variably penetrant disease necessitating the assessment of family history and a cautious, less intrusive approach to prevention. In contrast, BRCA mutations necessitate no less than surgery, betraying the belief among researchers that risk is functionally 100 percent. Developments in ovarian cancer detection and treatment—such as more and more calls for genetic testing and strongly worded recommendations for aggressive, sometimes experimental, procedures—suggest that in the practice of ovarian cancer research and treatment, heredity itself is the disease. Just as a doctor would not recommend surveillance instead of treatment for a woman with cancer, so too surveillance for a woman at risk is not considered to be a rational choice in the evolving discourse of ovarian cancer.[131]

As I have tried to show, the reinvention of medical practices is not necessarily a conscious decision on the part of researchers and physicians. Rather, new practices reflect the convergence of institutional interests and conceptual change, the mapping of which can be accomplished by an analysis of discourse. In the 1930s growing attention to ovarian cancer helped revitalize oophorectomy after reform efforts had successfully ended the procedure for reasons having to do with what was perceived to be noncompliance with gender norms and the pathologies of femininity more generally. Despite the rise of evidence-based medicine and the accomplishments of the feminist and women's health movement, oophorectomies are now being performed at an alarming rate. And although recent studies have prompted critical discussion about these rates, the development of BRCA tests promises to provide a secure place for oophorectomy in gynecological care.

But more than that, BRCA status offers new discursive means to gender the body and in turn justify particular types of interventions. Whereas physicians in the nineteenth century could explicitly draw on widely circulating beliefs about the relationship between reproductive organs, mental health, proper femininity, and motherhood, contemporary genomics has at its disposal the language of genetic risk, articulated via the language of expert knowledge and scientific objectivity.[132] This is not to say that gender norms no longer play a role in the constitutive function of scientific discourse. As the next section will show, BRCA research takes place within an established set of medical practices that in turn reflect particular understandings of gender, reproduction, and disease.

BRCA, Oophorectomy, and Gender

The history of oophorectomy shows, as does the recent controversy surrounding the Parker decision model, that certain fundamental assumptions about the role of ovaries in health and disease have gone largely unquestioned. Although ovaries are no longer removed for so-called psychic disorders, cancer prevention has replaced this earlier rationale for prophylactic removal. Similar to abusive practices of the past (for example, Battey's or Tait's operation), the current model of ovarian cancer prevention rests on unquestioned and/or flawed assumptions—today, those stemming from static, woefully underdeveloped theories of ovarian function[133] and the concomitant inability of medicine to model these functions effectively. Still, the recent debate about prophylactic oophorectomy for women understood to be at average risk for ovarian cancer inaugurates an important moment in this history: the studies by the Parker and Rocca teams, irrespective of methodological limitations, destabilize the foundation on which this procedure has heretofore been carried out. Despite the inertia of standardized, routinized medical procedure, there may be some chance that reflective examination of practices will continue in earnest, as well as the pursuit of much-needed research on the multifaceted function of ovaries throughout the life span.

Yet, as I discussed in the previous section, even within the pages of these articles and the responses to them, there remains one clearly stated exception to the findings: the BRCA mutation carrier. As Parker and colleagues make clear, their analysis pertains to women with average risk for ovarian cancer, not women with the higher risk associated with inherited mutations of the BRCA genes. Even with the increase in mortality associated with oophorectomy, the information that a BRCA test provides is sufficient to justify the surgery. On closer examination, however, many of the criticisms of routine prophylactic oophorectomy apply to BRCA carriers as well, thus inviting critical reflection about their putatively distinct medical status. The wide range of risk for ovarian cancer (10–60 percent) among BRCA carriers at the very least necessitates the assessment of the risks associated with surgical, early menopause. There are, moreover, considerations specific to BRCA carriers who contemplate prophylactic measures.

First, although oophorectomy significantly reduces the risk for ovarian cancer, it does not completely eliminate it. The studies by the Kauff and Rebbeck teams published in 2002 did show a dramatic decrease in ovarian cancer. However, for some women the procedure did not decrease risk for peritoneal cancer—a phenomenon that has been observed for some time and that proves what Andrea Eisen and Barbara Weber have called BRCA's "field" effect.[134] In a third study published in 2002, a different team of researchers concluded that "the majority of the ovarian-type cancers observed in our cohort were determined to be of peritoneal origin and, we believe, would not have been prevented through prophylactic salpingo-oophorectomy."[135]

Second, if BRCA mutation carriers begin HRT in order to compensate for the deleterious effects of premature menopause, they would be voluntarily taking a drug known to increase risk for breast cancer (recall that the BRCA genes are implicated in both breast and ovarian cancer).[136] These women could, perhaps, take estrogen-only HRT, although they would do so with the knowledge that this placed them at higher risk for uterine cancer.[137] The HRT conundrum has led researchers and practicing physicians to recommend that women have a complete hysterectomy (the removal of uterus, ovaries, and fallopian tubes), which normally requires major abdominal surgery. Relatively new methods such as laparoscopic-assisted vaginal hysterectomy (in which the uterus is removed through the vagina) present their own complications. According to one team of researchers, "the single biggest predictor of complications associated with laparoscopic procedures was the inclusion of hysterectomy as part of the procedure."[138] (Oophorectomy carries its own surgical risks: in one study, 4 out of 98 women experienced complications such as infected wounds, punctured bladders, punctured uteri, and bowel problems).[139]

As one might imagine, the decision to undergo prophylactic oophorectomy is a difficult one. A woman must calculate, on the one hand, the risk associated with her BRCA mutation and how much an oophorectomy will reduce that risk. On the other hand, she must calculate the risk of cancer after surgery, the risks associated with the procedure itself, and the health risks associated with premature menopause. In order to aid the decision-making process, experts have developed models for calculating the relative benefit of oophorectomy.

The information provided by models, however, is inextricably linked to the assumptions made by those who construct them, as can be seen in the earlier discussion of the Parker study (itself an attempt to correct for earlier decision models). The Grann decision analysis model, for example, assumes that a woman with a BRCA mutation will take Tamoxifen and that the drug will substantially reduce her risk for breast cancer.[140] But the evidence is mixed as to whether women with BRCA2 mutations would actually benefit from this form of chemoprevention.[141] The Grann model also assumes that since a BRCA mutation carrier is already at an exceptionally high risk for breast cancer, the administration of HRT would not—indeed, could not—increase her risk by an appreciable degree.[142]

This reasoning makes sense only for women who have an 85 percent lifetime risk for breast cancer (if it makes sense at all). It ignores the well-documented variability of BRCA-linked breast cancer risk, and it overlooks the fact that many women test positive for mutations for which no clinical trial evidence exists. The Grann model further assumes that the risk for heart disease in women with BRCA mutations is the same as in women in the general population, despite evidence suggesting that the risk may be higher for women experiencing surgical menopause. And finally, the model assumes that all women face the same risk for mortality from ovarian cancer, even though some studies show that women with BRCA mutations have a better chance of surviving that cancer than women without mutations.[143]

Even with the contestable assumptions of the Grann model, the greatest life expectancy gain predicted is 4.6 years for a thirty-year-old woman who undergoes oophorectomy and chemoprevention. A different model predicts more modest gains: 0.3–1.7 years for a thirty-year-old woman who chooses oophorectomy.[144] Neither model takes into consideration spotty HRT use over time or evidence that early, or premenopausal, surgical oophorectomy can be dangerous. Regarding the latter point, it is certainly not the case that the health benefits of intact ovaries are completely ignored by researchers. Nevertheless, the treatment decision calculus presented in studies of the relative benefit of oophorectomy does not take intact ovaries to be important enough to weigh against the benefits of cancer prevention. The short- and long-term side effects of oophorectomy often recede, rhetorically, in a discourse in which the only goal is reducing cancer risk, seemingly at any cost. In one influential study supporting the efficacy of oophorectomy, the researchers first construct a decision calculus that clearly downplays the side effects associated with oophorectomy:

> The primary negative consequence of prophylactic oophorectomy in premenopausal women is premature menopause, which may be associated with increased risks of osteoporosis and cardiovascular disease. Hot flashes, vaginal dryness, sexual dysfunction, sleep disturbances, and cognitive changes associated with menopause may affect the quality of life. However, the risk is balanced by the *morbidity and mortality* associated with breast and ovarian cancer in carriers of BRCA1 or BRCA2 mutations, and these symptoms may be managed by hormonal or nonhormonal medications.[145]

After strategically qualifying cancer risks with an explicit reference to mortality, the researchers conclude by recommending oophorectomy, in part by dismissing any chance that surveillance could detect ovarian cancer in BRCA mutation carriers:

> On the basis of our study, we advocate prophylactic oophorectomy to reduce the risk of ovarian and breast cancer in women with BRCA1 or BRCA2 mutations. In deciding whether to undergo the procedure, a woman should take into account how long she wishes to maintain fertility, and she should receive counseling about the risk and benefits of prophylactic oophorectomy. The decision should also be made with the knowledge that current surveillance regimens have not been shown to affect the incidence of late-stage ovarian cancer. Although opinion is divided on the use of hormone-replacement therapy after prophylactic oophorectomy, the decision to use estrogens should be based on a consideration of symptoms that affect future health and the quality of life. Some centers routinely recommend hormone-replacement therapy after prophylactic oophorectomy until the age of 50 years, and many women consider prophylactic oophorectomy unacceptable without this option.[146]

These statements reveal the implicit belief that a woman's fate is, after all, in her genes, and with new knowledge of BRCA, heredity is materially observable in ways that justify all feasible interventions. They are also expressions of the phenomenon of medical specialization that necessitates and, in turn, excuses myopic attention to discipline-specific, preferred objects of study. And they reflect the ways in which fear of cancer is as much a cultural artifact as it is material evidence of embodied experience in particular moments in history. Women are much more likely to fear breast cancer than they are heart disease, even though the latter exacts a far higher death toll than the former.[147] Some research suggests, in fact, that fear of cancer overdetermines some women's decision to undergo prophylactic oophorectomy.[148]

But more than this, the exceptional status that BRCA research occupies within the evolving debate about oophorectomy provides a new patient population for the procedure at the very time when its legitimacy is in crisis. Similar to the discovery of ovarian cancer as a justification for oophorectomy in the late 1930s, interest in the treatment of BRCA mutation carriers is a historical event in which a number of institutional interests have converged, creating a new constituency for the long-embattled procedure. Describing the material influence of BRCA testing on ovarian cancer prevention, Kenneth Offit, a well-known oncologist, said: "We are beginning to believe that we may be showing some first evidence of lives saved because of the genetic testing we are doing. . . We have now followed over 200 women with BRCA mutations, and we have found four early-stage ovarian cancers in almost 100 preventive surgeries. For many of us, it is exciting. I have been at Memorial for 10 years and didn't see a stage I ovarian cancer until we started doing BRCA testing. . . These developments have energized investigators to press ahead."[149] In a mutually beneficial turn of events, oophorectomy imbues the BRCA research agenda with a measure of legitimacy it had previously lacked. Widespread acceptance of the BRCA test has been hampered by the dearth of options available to women should they learn that they carry an inherited mutation. And of those options, most are radical, like prophylactic mastectomy, oophorectomy, and pharmaceutical treatments. One study noted: "Further information on the benefits of genetic testing might help women to make decisions about whether to have genetic testing, and will allow women to use information gained from testing in an optimum way that will improve clinical outcome."[150]

The legitimacy of prophylactic oophorectomy is further bolstered by its association with reduced risk for hereditary breast cancer.[151] The Rebbeck and Kauff teams concluded that prophylactic oophorectomy reduced not just ovarian cancer risk, but also breast cancer risk for the women they studied. Since then, several researchers have speculated that women would in fact prefer oophorectomy to mastectomy to avoid the disfiguring consequences of the latter.[152] The established efficacy of oophorectomy, moreover, has led to calls for more genetic testing as well as for including the surgery as an option in counseling women at risk for breast cancer whether or not they have inherited a BRCA mutation. One physician wrote: "The effectiveness of prophylactic oophorectomy in carriers of BRCA mutations provides a strong rationale for genetic testing in women with a strong family history of breast cancer."[153] And in 2001 researchers from the United Kingdom advised: "Epidemiological studies suggest that prophylactic oophorectomy may play a role in the prevention of breast cancer in high risk populations. Therefore, there may be a case for informing women at risk of breast cancer about this option."[154] So strong is the suspected benefit of prophylactic oophorectomy for reducing breast cancer risk that it is slowly becoming a thinkable strategy for the prevention of breast cancer in groups other than BRCA mutation carriers. In a review of the literature, Noah Kauff and Richard Barakat conclude that even in families in which no BRCA mutations are detected, women might still benefit from prophylactic oophorectomy—at least one study, they say, shows that "RRSO [risk-reducing salpingo-oophorectomy] is protective against breast cancer at all levels of risk."[155]

This is not the first time that the relationship between hormones and breast cancer has resulted in the endorsement of radical approaches to the disease's treatment and prevention. Indeed, "salpingo-oophorectomy has a much longer history in the prevention and treatment of breast cancer than in the prevention of ovarian cancer."[156] It was first proposed as a treatment for breast cancer in 1889 and first performed for this purpose seven years later. In 1950 A. M. Liber suggested prophylactic oophorectomy as standard of care for breast cancer.[157] Breast cancer treatment and prevention have included even more extreme measures, including adrenalectomy, the removal of all of a woman's adrenal glands. Charles Huggins and Thomas Dao, authors of a 1953 study of breast cancer, reasoned that because hormones were detected in the urine of women after oophorectomy, removal of the adrenal glands would ensure abrupt and total cessation of her body's hormone production. To compensate for what they audaciously describe as "an easily manageable deficiency of adrenal glands," women in this study were treated with cortisone and desoxycorticosterone, presumably for the rest of their lives.[158]

Over the years, radical mastectomy would become doctors' preferred method of treating and preventing breast cancer, because it did not require long-term steroidal and hormonal treatments, did not increase risk for cancer and other diseases, and did not result in infertility. Moreover, studies of oophorectomy for breast cancer prevention cast doubt on its efficacy when researchers observed that it did not provide any advantage in terms of recurrence or survival.[159] In the 1960s, says Sharon Batt, oophorectomy as a breast cancer treatment fell out of favor.[160] How, then, can it be that oophorectomy is once again a thinkable approach to the treatment and prevention of breast cancer for women with a family history of the disease?

Arguably, recently published landmark studies correct for the insufficient evidence that oophorectomy can decrease breast cancer risk. Furthermore, oncologists' ability to distinguish between breast cancer tumors that are sensitive to estrogen and those that are not, together with the subsequent popularity of chemoprevention drugs that regulate estrogen production, no doubt make the manipulation of the endocrine system an attractive approach to treatment. Important, too, is the influence of women's health advocacy. The disfiguring effects of radical mastectomy, especially the version popularized by William Halsted, catalyzed the modern breast cancer movement and the eventual decline of both the radical mastectomy and the surgery's other varieties. Nevertheless, although lumpectomy has largely replaced mastectomy for women in the general population diagnosed with breast cancer,[161] prophylactic mastectomy is recommended for women with BRCA mutations,[162] and so alternatives to mastectomy are still a topic of concern for this group of women.[163] In their evaluation of statistics regarding prophylactic mastectomy and prophylactic oophorectomy, Andrea Eisen and coauthors claim that the numbers reveal more interest in the latter, due to the "absence of externally visible physical changes."[164]

The emergence of oophorectomy for the treatment and prevention of breast cancer provides, in serendipitous fashion, a new rationale for the surgery at a time when its legitimacy is being challenged in other areas of gynecological medicine. BRCA genomics in turn (and genomics more generally) requires evidence-backed interventions that will help make genetic testing more attractive to women at risk. What the literature on oophorectomy and breast cancer also reveals, however, is how gender norms both shape and are shaped by the discourse of heredity and the female body. As this chapter has already shown, the metonymical reduction of ovaries to their reproductive function has been absolutely key to the exception made for BRCA mutation carriers. When cancer researchers effectively dismiss the health importance of ovaries as well as the complexities that HRT presents for these women, they presume that ovaries enable procreation but nothing else beneficial or productive. Thus, ovaries are dispensable for women who no longer need or want them for the purpose of reproduction.

In the literature on oophorectomy and breast cancer, sex is similarly gendered. Ovaries are clearly distinguished from breasts when one accounts for the preference by women and their doctors for one surgery but not another. The distinction is only possible because of the presumed difference between visible and invisible sex organs (or, one might say, the difference between sex organs and reproductive organs). Statements regarding the superiority (aesthetically speaking) of oophorectomy imply that breasts are visible markers of femininity and sexuality in ways that ovaries are not. Invisible to the naked eye, ovaries are not part of the gendered economy of signs in the way that breasts are.[165]

One conclusion that could be drawn from this observation is that one marker of sexuality and femininity, breasts, are considered more worthy of preserving than another marker, ovaries. This, however, is not the case. Fertility—and, by extension, motherhood—remains a privileged experience of embodied femininity precisely because of its relation to cancer risk. What is essentially the link between fecundity and cancer was forged long before discussion of oophorectomy for BRCA mutation carriers proliferated in the cancer research literature. According to the so-called estrogen hypothesis, the more estrogen a woman is exposed to in her lifetime, the higher her risk for cancer. An increase in parity (the number of childbirths) can decrease breast cancer risk, because estrogen levels drop considerably when a woman is pregnant. Pregnancy and childbirth at a young age have an even greater protective effect, not only because of the shorter duration of estrogen exposure over the course of a woman's life but also because pregnancy induces breast cell differentiation, thereby reducing the cells' susceptibility to mutagens. Parity can also decrease risk for ovarian cancer because pregnancy reduces the number of ovulation cycles and, consequently, the chance that ovarian cells will mutate.[166] Motherhood, it seems, exerts an unquestionably protective effect and so occupies, if only indirectly, a position of privilege in biomedicine, a discursive phenomenon that can only reinforce its privileged status in the broader culture.

For women at risk for ovarian cancer, the politics of reproduction (and, by extension, motherhood) play a central role in prevention discourse, because oophorectomy results in infertility. Cancer researchers, as a consequence, have to negotiate how cancer risk management affects the reproductive choices available to women. Recommendations carefully qualify the timing of oophorectomy by expressly advising that it occur soon after "childbearing is complete."[167] So in the effort to protect reproductive autonomy, researchers recognize, if only implicitly, that fear of cancer should not interfere with the desire for motherhood. According to the logic of genomics-inflected cancer prevention guidelines, for women both capable and desirous of motherhood, cancer risk management can and should be balanced against the decision to start a family.

The motherhood exception thus satisfies the bioethical stipulation that medical technology respect, and protect, women's interests. Motherhood, however, appears to be the only exception to the norm—there seems to be no other acceptable reason for postponing oophorectomy or avoiding it altogether. For example, a BRCA2 carrier, who will probably be diagnosed with ovarian cancer at the same age as women in the general population, may conclude that risk for heart disease and osteoporosis weighs heavily against the benefits of early oophorectomy.[168] The rhetoric of timing, moreover, determines whether a woman can in fact exercise motherhood as an exceptional state, and it does so by trafficking in mainstream, accepted gender norms. If, for instance, a woman contemplates postponing oophorectomy beyond the age of thirty-five, her physician will have to consult a medical literature that has already determined such a risk to be unacceptable, especially if the woman in question is a BRCA1 mutation carrier (recall that women with BRCA1 mutations tend to be diagnosed six to nine years earlier than women with BRCA2 mutations).[169] This disparity in the timing of cancer diagnoses "has significant implications," the authors of one study argue, "for counseling BRCA mutation carriers regarding the timing of preventative interventions."[170] For BRCA1 mutation carriers, early motherhood is necessary in order to reap the benefits of prophylactic surgery (although, interestingly, BRCA2 mutation carriers are never presumed to have good reason to postpone oophorectomy beyond the age of thirty-five, or to refuse it entirely). In short, reproductive options cannot be fully retained without incurring some health risks:

> With regard to the question of timing, there is uniform consensus, *for obvious reasons*, that risk-reducing salpingo-oophorectomy should be deferred until childbearing is complete. However, as more of our patients delay child bearing into their late 30s and 40s, BRCA1 mutation carriers, in particular, begin to expose themselves to a nontrivial rate of ovarian cancer (11–21 percent by age 50). Although carriers of BRCA2 mutations are at less risk of ovarian cancer during the reproductive years, mutation carriers who defer oophorectomy until the 40s and beyond may lose the substantial protective effect of salpingo-oophorectomy against breast cancer [emphasis added].[171]

Presumably, cancer researchers have chosen the otherwise arbitrary age of thirty-five because, in the field of obstetrics, it is the number that separates young from so-called advanced maternal age.[172] Yet fertility trends show that women are having children later in life, often well into their thirties and early forties.[173] Within this context, to say that age thirty-five is the cutoff for childbearing is to say that it ought to occur relatively early in life.

There are many reasons why women delay childbearing, including but not limited to pursuing an education, a career, or financial security. Women have been told for some time that this decision, whether or not of their own making, has psychological costs, described by Susan Faludi in *Backlash*.[174] If the estrogen hypothesis is true, that cost also includes an increased risk for breast cancer. And for women with BRCA mutations at risk for ovarian cancer, postponing oophorectomy risks near-certain diagnosis of a very deadly disease—a diagnosis that is avoidable, given the presumed protective effect of oophorectomy. The discourse of inherited ovarian cancer, then, bears traces of early-twentieth-century theories of reproductive organs and well-being discussed above in the chapter, although for obvious reasons the association of motherhood with disease has not remained as explicit or direct. Nevertheless, ovarian cancer risk discourse is the most recent version of a cultural and medical logic that holds that when a woman resists traditional norms of femininity and sexuality, her unproductive ovaries become the source of an impressive array of health disorders.[175]

The synecdochical reduction of women to ovaries, and then of ovaries to reproduction, is linked in part with other institutional discourses through which women are hailed as mothers. Take the example of employment law. Women have experienced workplace discrimination for the putative purpose of protecting and preserving their fertility. In the *Johnson Controls* case, women workers sued to protect jobs in industrial settings in response to company policies banning fertile women from positions that exposed them to hazardous chemicals and their employers to lawsuits. These litigants challenged the presumption that women's lifeworld should be reduced to their presumed capacity to reproduce.[176]

Yet even when women challenge the ontological conflation of sex and gender, most notably with the feminist argument that women should not be confined or limited in any way by biological difference, sex nevertheless seems to materialize gender norms.[177] Even the social construction argument—as successful as it has been in challenging, to some degree, the enforcement of strict gender roles—presumed, although by default, that only gender is socially constructed. But, as Judith Butler countered, to assume a distinction between biological matter and culture is to ignore the very powerful way in which the body exerts a disciplinary effect precisely because it is a signifier of proper gender roles:

> If the immutable character of sex is contested, perhaps this construct called "sex" is as culturally constructed as gender; indeed, perhaps it

was always already gender, with the consequence that the distinction
between sex and gender turns out to be no distinction at all. It would
make no sense, then, to define gender as the cultural interpretation of
sex, if sex itself is a gendered category. Gender ought not to be con-
ceived merely as the cultural inscription of meaning on a pregiven sex
(a juridical conception); gender must also designate the very apparatus
of production whereby the sexes themselves are established. As a result,
gender is not to culture as sex is to nature; gender is also the discursive/
cultural means by which "sexed nature" or "a natural sex" is produced
and established as "prediscursive," prior to culture, a politically neutral
surface *on which* culture acts.[178]

Reproductive organs ("sex") serve to gender bodies materially and socially; reproduction, says Butler, has become the only salient feature of
sex, even though there are many bodies that resist or simply exist outside this grid of intelligibility.[179] In the present case, ovaries are the site
of the imposition of gender norms, a disciplinary effect of culture that is
masked by the power of biomedical discourse to present the biological
body as natural fact. If the biological body is always already gendered, it
stands to reason that reproduction would be the sole exception to oophorectomy. Ovaries are engines of procreation or creeping death, and this
binary governs the body in two ways: first, by tacitly enforcing norms of
motherhood (in particular, young motherhood); and second, by securing the body as an object of surveillance and intervention by the field of
cancer medicine.

Although it is not possible to live wholly outside of such discursive constraints, it is nevertheless the case that the enactment of gender norms
is not entirely successful all of the time. Some bodies are not fully intelligible according to the logics of normative sex or gender and so could be
said to challenge or resist them in some way. These bodies provide an occasion—if only a fleeting one—to comprehend the performativity of the
gendered body: the supposed failure to repeat, or properly enact, gender
opens up the possibility for questioning its biological (or even metaphysical) substance. The intersex are a case in point: the intersex body defies
the traditional male and female sex categories, whether they are defined
by chromosomes, hormones, genitalia, or reproductive organs.

With the recognition among feminist social theorists and science studies scholars that sex is no more a biological fact than gender is, even
though the successful enactment of gender provides the illusion that it is,

the social (and even ontological) status of the ovary is arguably in flux. In the aftermath of the Parker and Rocca studies, so, too, is its medical status. The bodies in these studies, by providing evidence of ovarian function beyond reproduction, refuse to conform to accepted biomedical theory of the female (that is, feminine) body. And women who postpone or refuse prophylactic oophorectomy—and not just for reproductive purposes—transform their bodies into sites of resistance to accepted medical practice, bodies that are virtually unintelligible in a biomedical discourse that equates such refusal with death.

In response to unruly bodies, such as those of the intersex, physicians and parents have, more often than not, opted to enforce gender norms through invasive means, including surgery and intense psychological therapy, violence sanctioned because of the power of gender to elicit compliance with its dictates. There simply is no language for comprehending these bodies. Even the word "intersex" shows the difficulty in representing certain bodies within language, as it presumes a third sex, somewhere in between male and female, both of which remain present—if only as a trace—in the very word "intersex."[180] The genomics discourse I have examined in this chapter tethers ovaries to reproductive function and desire by making motherhood the only acceptable reason for postponing oophorectomy. In so doing, the discourse disciplines these bodies at the very time that feminist theory seeks to destabilize hegemonic gender norms.

The production of the female medical subject of genomics imputes legitimacy to a procedure increasingly under scrutiny and to a genomics research program struggling to integrate itself into routine medical care; and, as I have argued, this medical subject must also be aligned with widely held beliefs about reproduction and the normative body. What emerges is a calculus wherein the exercise of reproductive autonomy is increasingly a risk to one's health over time. Women with BRCA mutations, then, exercise reproductive choice (or defy it altogether) within a constrained field of risk. The reproductive body of genomics thus serves institutional interests while simultaneously aligning itself with cultural beliefs about the figurative and material costs of feminism.

The Racialized Reproductive Body

Although the scientific discourse described in the previous section does not distinguish between white and brown bodies, one is nevertheless struck by the apparent contradiction that inheres in the privileging of young motherhood and the pathologizing of the same (and, indeed, of motherhood in general) for African American women within the larger culture.[181] Indeed, throughout history discourses of motherhood and feminity have been structured around the presumption of racial difference. Put another way, they are racialized. Early twentieth-century diseasing of white women's reproductive organs was possible because of the belief that the exercise of their rights compromised reproductive capacity and thus posed a threat to the social order, culminating in eugenicist fears of race suicide. In contrast, in the case of black women, the source of such threats was their presumed hypersexuality and fertility. Whatever the paradoxical relation oophorectomy had to this differential cultural imperative to reproduce, the importance of white fertility was never in doubt and, indeed, provided a narrative that critics of the surgery could appeal to during nineteenth-century debates.

As Dorothy Roberts and Rickie Solinger have both documented, policymakers and public intellectuals of all political persuasions have typically directed their ire toward black women, rather than institutional racism, when trying to explain phenomena such as poverty and crime.[182] What has resulted are differential practices of hysterectomy and sterilization along the axis of race.[183] As discussed in chapter 2, sterilization was a popular surgery beginning in the early twentieth century, escalating in the 1920s with the rise of negative or mainline eugenics. Although the demise of the eugenics movement reduced the number of sterilizations performed, involuntary and coerced sterilization of US black women continued well into the 1970s.[184] Moreover, implantable and injectible contraceptives such as Norplant and Depo-Provera effectively sterilize women because they induce long periods of infertility and, in the case of implantables, require costly medical visits to remove.[185] Today, sterilization is one of the most common forms of birth control for African American women in the United States.[186] And although hysterectomy rates have been artificially high for women across class and race, they are particularly high for black women.

When the ovaries of black women symbolize reproductive capacity, they are targeted for removal. When they are cancerous and pose grave threats to life, they no longer serve as objects of radical intervention—a reverse of the logic I described in the previous section, when, presumably, the bodies in question are white. Regarding gynecological cancer, differences in mortality from ovarian cancer are explained in part by the fact that African American women are more likely than white women to be diagnosed at an advanced stage of the disease and in part by the differential treatment black women receive from their doctors,[187] and this is true even when adjusted for age and for disease stage at the time of diagnosis.[188] Specifically, surgery is less likely. According to Groesbeck Parham and coauthors:

> African-American women with advanced invasive epithelial ovarian carcinoma were less often treated with combined surgery and chemotherapy and more often treated with chemotherapy only. African-American women were twice as likely as white women not to receive appropriate treatment. African-American women had poorer survival rates than white women from the same or different hospitals, regardless of income. Among staged cases, African-American women were more often diagnosed with Stage IV disease than either group of white women.[189]

Another report on the related topic of endometrial cancer notes:

> The results of this study demonstrate that African-American women with endometrial cancer are less likely to undergo primary surgery and are more likely to die from their disease. Furthermore, in our study, this racial disparity was observed for women with both local-regional (stages I, II, III) and metastatic (stage IV) disease. Furthermore, because surgery is significantly associated with survival for both these groups, our study demonstrates that racial differences in survival are smaller after adjusting for lower rates of surgery among African-American women.[190]

These statistics show significant differences in mortality from gynecological cancers, the result of racism in medicine. Yet one would expect breast cancer rates to be lower for blacks than for whites, especially for those black women who have multiple children and start having them early in life (if stereotypes are to be believed). As discussed above, reproductive history, aside from exposure to ionizing radiation, is the only virtually undisputed risk factor for breast and ovarian cancer (having fewer cycles of ovulation reduces exposure to estrogen and lowers the rate of cellular division in ovarian cells, and during pregnancy breast cells become more developed and distinct—more stable—and therefore less suscepbile to mutagenesis). Yet rates of breast cancer among this group of black women are high, especially in comparison to rates among white women.

One explanation for this discrepancy appeared in a 2003 article in the *Journal of the National Cancer Institute*, in which the authors suggested that when young women become pregnant, a window of vulnerability opens that may make them more susceptible to malignancies.[191] In this way, the study seemed to lend credence to Nancy Krieger's 1989 critique of the estrogen hypothesis, in which she argued that many patterns of breast cancer incidence contradict the putative impact of endogenous hormones like estrogen and that exposure to external carcinogens at particular times during the development of the breast probably explains much breast cancer incidence.[192] The findings of the 2003 study ostensibly apply to all women, yet the authors were especially interested in how their theory might explain why, when fertility rates are higher among African American women, their breast cancer rate is higher, not lower, in contradiction to the estrogen hypothesis. The implication of the study, beyond its scientific claims about the relationship between parity and breast cancer risk, is that pregnancy for young black women is a life-risking enterprise. (The study also feeds the myth of higher fertility rates among African American women; the authors assert that rates were higher at the time of their study, although their data were from 1998. Even if fertility rates were higher for the group of women they studied, it was nevertheless the case that rates in 2003 were much lower than the study's authors acknowledged.) Indeed, the study cannot account for the present state of affairs: fertility rates are the same between young black and white women, yet breast cancer rates are significantly higher for the former.[193]

This, then, is the context in which genomics risk discourse takes form, a context in which the intersection of gender and race appears to have particular significance. If the history of racism in medical practice provides any lessons, we can expect that an African American woman at risk will probably not have the same access to cancer prevention techniques as a white woman, including prophylactic surgery. At the same time, the ovaries of black women are considerably more vulnerable if institutional interests lean toward their removal. Genomics, as it turns out, is particularly dependent on the racialization of mutations in order to extend its reach and to maintain its legitimacy in the eyes of funding institutions, a topic I discuss in chapter 4. The population turn in genomics permits the development of screening techniques based largely on skin color and presumed ancestry rather than on family history, the latter being a much more delimiting criterion.

Yet BRCA screening has met considerable resistance within the African American community.[194] There are a number of observed reasons for this (lack of understanding of genetic testing, fear of genetic discrimination and social stigma, distrust of the medical community, low perception of risk and/or concern about risk, and other variables such as religion and family dynamics), but the cumulative result is that testing rates are much lower among black women than white women.[195] One study calculated that African American women are 78 percent less likely to use genetic counseling than white women.[196] Moreover, even when interest in genetics is high, the actual use of genetic services (even when free) does not seem to correlate with that interest.[197] When testing is for research purposes only, participation rates are especially low.[198]

The result has been a persistent, urgent call—at times framed as an issue of social justice and health disparities—for greater outreach in the black community. There is a legitimate need for greater understanding of the diversity of the human genome and of the BRCA genes in particular. Not only does low uptake of testing undermine BRCA's relevance to a significant proportion of the population, but it also ensures the continuation of this state of affairs because not much is known about the mutations likely to be detected among this group of women. Outreach efforts betray the belief that the kinds of preventive measures available to women, such as prophylactic oophorectomy, should be available to all women, regardless of race or class. In fact, the accretion of evidence regarding this surgery's efficacy provides a "stronger rationale" for genetic testing and counseling.[199] As more black women are persuaded to undergo testing, and as insurance companies add prophylactic surgery to the list of covered procedures, the number of sterilizations in the name of managing inherited risk may rise. Whether this outcome will be an extension of abusive gynecological practices of the past remains to be seen.

When an association between BRCA mutations and hereditary forms of ovarian cancer was made early in this research's history, it seemed as though a desperately needed tool in the fight against a highly lethal disease had been discovered. And although the available course of action—early surgical removal of a woman's ovaries—is far from ideal, the deadliness of ovarian cancer invites a risk assessment that compels the adoption of any and all measures to reduce risk.

Yet when we critically reflect on the science of inherited ovarian cancer as a positivistic discourse that masks its own embeddedness in history and culture, a more complicated narrative emerges, one in which BRCA research discovers new ways to enhance its legitimacy while simultaneously providing cover for the field of gynecology as it finally comes to terms with its own legacy of misogyny—a legacy that still informs its method of preventing cancer: prophylactic oophorectomy for any woman for whom the surgery can easily and readily be performed. Indeed, both cancer genomics and gynecology depend on fear of ovarian cancer to deflect critical attention from their conceptual and material practices.

The danger of ovarian cancer is real, but it is also relative: surgery does largely reduce risk for ovarian and similar cancers, but it does so at the risk of fatal heart disease, osteoporosis, and brain disorders. The fact that the discourse of ovarian cancer can absorb these other risks is less a matter of its facticity than of its emergence as a mode of legitimation for gynecology in the early twentieth century. Although the surface-level justifications for prophylactic oophorectomy may have changed from the treatment of psychic disorders to the reduction of cancer risk, the justificatory logics are similar, as is the impact on women's bodies.

The recent research on prophylactic oophorectomy has shaken the foundation on which this surgery is routinely and uncritically performed. Yet at the same time, the effectiveness of oophorectomy in reducing risk for BRCA mutation carriers means that the surgery is a crucial mode of legitimation both for gynecology and for genomics. We have seen how, when the risks for BRCA carriers are the same as those for noncarriers, prophylactic oophorectomy remains relatively uncontroversial: the exceptional status for BRCA mutation carriers is possible because of the equation of heredity with fate. To carry a BRCA mutation is, materially speaking, the same as a diagnosis of cancer, for which nothing less than radical surgery is a thinkable mode of action.

A rethinking of oophorectomy for women in the general population has entailed resignifying ovaries as agents of health and well-being, and not merely as agents of procreation. But BRCA discourse reverses these critical rhetorical and theoretical developments, first by diseasing ovaries peripherally through the identification of pathological mutations, then by designating motherhood as the only acceptable reason to avoid surgery (albeit within certain age-related constraints). Inherited ovarian cancer discourse is dependent on the dominant, common-sense characteristics of the reproductive body (women have the right to fulfill their procreative desires; those desires ought to be satisfied by the middle of their third decade of life) while it also alters motherhood's comprehensibility due to the dictates of genomics' own knowledge claims (because of cancer risk, motherhood must be practiced within a field of constraint). BRCA research, then, is a recuperative project, both for its own ever-changing institutional needs and for a larger social order threatened by feminist and queer theorizing that destabilizes our unquestioned beliefs about the biological foundations of the sexed and gendered body.

Although bodies are irreducibly gendered in ideological discourse, they are also racially marked. The racial politics of reproduction structures scientific discourse differentially. BRCA and ovarian cancer discourse develops amid the paradox of hysterectomy and sterilization abuse carried out alongside the undertreatment for cancer and underutilization of BRCA screening among black women. The aggressiveness with which black women's healthy reproductive organs are removed by gynecologists means that this group's relatively marginal status in the world of genomics provides unexpected yet fortuitous protection from oncologists bewitched by the grand promises of BRCA research. At the same time, African American women's lower likelihood of receiving adequate treatment for ovarian cancer means that prophylactic oophorectomy for those at high risk might act as a counterweight to health disparities should geneticists successfully recruit black women as genetic screening patients. It remains to be seen what the outcome of aggressive outreach will be, but what is certain is that genomics researchers have a powerful argument to make, grounded in new evidence that oophorectomy is effective, that BRCA mutations confer unacceptable risk according to the logic of risk assessment in genetic counseling, and that the failure of genomics to reach diverse populations of persons is just one more instance of racial disparities in medicine. BRCA may negate the tendency to undertreat black women, but in a context in which the black reproductive body is, more often than not, subject to discipline and subjugation.

4

Genomics and the Racial Body

In 2006 the *New York Times* published an article by Denise Grady titled "Racial Component Is Found in Lethal Breast Cancer." The article opens with Grady's summary of research published in the *Journal of the American Medical Association*: "Young black women with breast cancer are more prone than whites or older blacks to develop a type of tumor with genetic traits that make it especially deadly or hard to treat, a study has found." This story, and the research it concerns, raises fairly serious questions about the meaning of race in medicine and in the broader culture. What does it mean for a "racial" group to be "more prone" to getting a particular type of cancer? What sort of "genetic traits" would differentiate tumors and, by extension, the women who have them? Should this research be valued for its contribution to understanding the significant difference in mortality rates between black and white women? Or does it merely perpetuate the popular, and dangerous, belief that race is materially real? Indeed, why is the word "race" used in a story about biological differences at all?[1]

The study's authors admit that the differences observed are what epidemiologists call "multifactorial" and so can be traced to such things as socioeconomic status and access to medical care. The focus of their study, the supposedly biological differences in the tumors themselves, they say can be attributed to genetic factors, lifestyle, and environmental exposures. They do not define "race" but simply say that self-identification of the tissue donors determined the groupings. For those study subjects who did not identify themselves as African American, the researchers used the label "non–African American"—even for women identifying themselves as members of other racial and ethnic groups, including whites. None of the black women in the study tested positive for BRCA1 mutations, despite the widespread belief that the kinds of breast tumors typical for this group of women are also typical for BRCA mutation carriers as well.

Grady's *New York Times* article addresses none of these complexities but instead describes evidence of tumor difference as a "genetic" discovery—ostensibly a hereditary one, since Grady interviews Olufunmilayo Olopade from the University of Chicago, who concludes that, in light of the study, African American women should consider genetic counseling if they have a family history of breast cancer. The article as a whole implies, then, that biological differences in the tumors of black women are the result of inborn traits.

This particular episode in the contemporary discussion of race and medicine reveals a number of important themes that this chapter will address. First, it shows that although disciplines vary with regard to how racial categories are employed to recruit, sort, and make risk claims about research subjects, what nevertheless emerges is a field-invariant understanding of race as a biological concept. The persistence of biological race is not merely the product of careless methodological practices on the part of scientists nor of careless reporting by science journalists. Rather, it suggests that certain habits of mind are circulating and manifesting themselves in discursive practices regardless of research questions, disciplinary norms, and linguistic conventions.

Second, the chapter shows the importance of studying race in the context of a specific disease. Although there has been a great deal of scholarship on the problem of race in genomics research, we must look at what happens when it is carried out in the context of a highly visible, common, and politicized disease. Indeed, when Grady was criticized for perpetuating the belief that "race" exists and can explain breast cancer disparities better than, say, the lived experience of embodied racial categories, she responded with the all-too-familiar pragmatism that characterizes Western medicine: if the use of race can save lives, it ought to be an operative concept in cancer risk assessment, at least until more refined categorization schema become available. Hence, the paradox we see in biomedical discourse of the simultaneous disavowal and recuperation of race. Race may be an invention and unfortunate legacy of slavery, but it is also our best hope for saving lives.[2]

In what follows, I describe the rhetorical phenomenon whereby health disparities are recast as hereditarian phenomena, thus securing genomics' rightful place in this growing area of research but also reifying race in the process. The phenomenon is rhetorical insofar as conceptual and linguistic displacements (themselves connected to methodological changes) mask the ways in which genomics discourse constructs racial subjects. I consider, for the purposes of the argument here, that genomics' racializing discourse is ideological insofar as it helps advance and secure state interests—specifically, the disinvestment in antiracism initiatives that would entail fairly significant redistribution of material resources. Indeed, the construction of race in genomics is important in a society that is, according to a number of indices, racist.[3]

A guiding assumption of the analysis I offer here is that a logic of progression conceals the ways that genomics performs important ideological work without explicit ideological appeals. It is thus an example of what Barbara Fields has described as the ideology of race, the language of which must explain the persistence of oppressive social relations, but that must also change to accommodate the changing face of those social relations. Racial ideology is a vocabulary that "need not and cannot be a duplicate of the one spoken by the rulers."[4]

The pathologization of black bodies, through the construction of breast cancer risk, not only affects the medical care and everyday lives of African American women, it also excuses a number of institutional practices that are responsible for the greater chance these women have of dying from breast cancer in the first place. The geneticization of African American women's breast cancer is, in many ways, the criminalization of black women's bodies—with testing-related stigmatization and discrimination and the administration of potentially dangerous preventive measures like surgery and drugs—and the decriminalization of the state's refusal to ameliorate the health effects of racialized social stratification.

For clarification purposes, I assume throughout this chapter that "race" is a social designation and that its biological significance inheres in the lived experience of race—the embodiment of social identity—that is in turn related to particular health outcomes. When I employ the terms "black" and "African American," it is to denote those persons identified and/or self-identified with these terms and who thereby occupy particular positions in a social and economic hierarchy.

The Population Turn in Genomics

Those familiar with popular discourse about the Human Genome Project (HGP) have heard Francis Collins and other advocates often count as one of its greatest achievements the production of evidence that all humanity shares virtually the same genetic makeup. When Collins and Craig Venter of Celera Genomics announced jointly that they completed a "rough draft" of the human genome sequence, President Clinton declared to the nation:

> All of us are created equal, entitled to equal treatment under the law. After all, I believe one of the great truths to emerge from this triumphant expedition inside the human genome is that in genetic terms, all human beings, regardless of race, are more than 99.9 percent the same. What that means is that modern science has confirmed what we first learned from ancient faiths. The most important fact of life on this Earth is our common humanity. My greatest wish on this day for the ages is that this incandescent truth will always guide our actions as we continue to march forth in this, the greatest age of discovery ever known.[5]

Yet despite the proof of common humanity that technology has provided us, it is the 0.1 percent variation of the genome that animates post-HGP research. Genetic variation, however small its share of the genome as a whole, is medically and, by extension, financially important. BRCA research is one such site of this trajectory in genomics research, because variants of both BRCA1 and BRCA2 qualify as risks for breast cancer—and there are, at last count, at least 3,500 of those variants.

Initially, researchers thought it unlikely that a single variant would be shared among women from different cancer syndrome families. In the early 1990s, however, geneticists discovered just such a mutation. Although the women involved were not from the same family, all were of Ashkenazi descent. This mutation was the first population-specific mutation of BRCA1, labeled 185delAG.[6] Moreover, the mutation was evidence of the so-called founder effect: the concentration of some mutations within a population, theoretically traceable to one ancestor. Subsequent to this finding, a research team in Washington, D.C., led by Jeffery Struewing, analyzed stored tissue samples from Tay-Sachs screening programs to test the hypothesis that the mutation was prevalent among Ashkenazi Jews.[7] The mutation was observed in 1 percent of the samples, a statistically significant percentage compared to the general population.[8] The researchers concluded that "the recognition of an unexpectedly high frequency of the 185delAG BRCA1 mutation suggests, perhaps for the first time, the realistic possibility of genetic screening of an adult population for predisposition testing for diseases with some potential for prevention."[9] These findings were confirmed two years later after the researchers worked with communities in the Washington area and recruited 5,000 participants for a follow-up study that examined not only the frequency of the mutation but whether, and to what extent, it increased risk for breast cancer. In 1997 Struewing and coauthors reported that the mutation was found with the same frequency as in their earlier study, and that it was associated with a higher risk for breast cancer. Interestingly, the risk was not as high as with other mutations: women with the 185delAG mutation were estimated to have a lifetime risk for breast cancer of 50 percent, belying the widely held popular belief that BRCA-related risk was not mutation specific.[10] In all, three mutations were identified as founder mutations in women of Ashkenazi descent.

By 1997 science writers were reporting the identification of many more founder mutations, conveying excitement among researchers that the population turn had firmly taken hold. An article in the *Journal of the National Cancer Institute* reported:

> Scientists had hoped the discovery in Ashkenazi Jews (those of Eastern or Central European descent) would help open some doors: to the development of inexpensive tests to check women from this ethnic group for an inherited susceptibility to breast cancer, and to the launch of population-based studies to further understand how BRCA1 mutations cause breast cancer. But the discovery of common mutations among other ethnic groups throws these doors wide open, raising the possibility of genetic tests for populations besides Ashkenazi Jews and expanding the horizons of population-based breast cancer research.[11]

In the United States, one of those research horizons would include African Americans as genomics researchers set out to describe the existence of founder mutations in black women. The working hypothesis was that if founder mutations existed, like those identified in the Ashkenazim, then perhaps other groups bound by ancestry rather than family would harbor BRCA mutations at a rate warranting targeted research and testing.

Three years after the BRCA genes were first described, in a research letter published in the *American Journal of Human Genetics*, Qing Gao and colleagues made the case for the study of African American breast cancer families. The study of such families, they argued, was justified given the fact that several earlier studies had included black subjects, some of whom had tested positive for BRCA1 mutations. However, in those previous studies, the number of African American study subjects was too small to be statistically significant. Gao and coauthors prefaced their remarks by observing: "Little information exists regarding BRCA1 and BRCA2 mutations in ethnic[12] groups other than Caucasians of northern European ancestry."[13]

In an effort to document BRCA mutations in black women in greater detail and with greater statistical significance, Gao and coauthors studied BRCA1 mutations in nine African American families contacted through the University of Chicago Cancer Risk Clinic. According to the researchers, the findings were significant. Five of the nine women screened (one woman from each of nine families) tested positive for mutations of BRCA. The researchers deemed three of those mutations "novel," as they had not previously been reported by other researchers (mutations are reported to a central database, called the Breast Cancer Information Core). "The majority of the cases listed in the database," Gao and coauthors observed, "are of northern European ancestry, and these mutations may not have been seen previously because they are exclusive to African Americans."[14] In addition to the two novel mutations designated deleterious (because they prevent the gene from working properly), two additional mutations were described as having "unknown significance"[15] because it was unclear how, or even if, they would impede the proper functioning of the gene. No mutations of any kind were detected in two of the families, including one family in which ten members had been diagnosed with breast cancer. And none of the "previously described African American mutations" was seen in this study.[16]

Despite its small sample size and the genetic heterogeneity it described, the study by Gao's team would set the stage for dozens more publications about African women; African American women with or without a family history of breast cancer; those with BRCA2 mutations; and those with more common, less damaging variations than autosomal dominant mutations of the BRCA genes. The impressive expansion of this area of BRCA research was possible for many reasons. First, the Gao team's study confirmed that black women in the United States do test positive for BRCA mutations.[17] In the words of its authors, "this study indicates that genetic susceptibility to breast cancer in this limited data set can be explained by BRCA1 gene mutations in 56 percent of high-risk African American families ascertained through young breast cancer cases referred to an urban cancer-risk clinic."[18] Second, it confirmed that the mutations detected are often novel, in the sense that, more often than not, they have not previously been reported—thus demonstrating the need to study African American women for whom extant research data are not necessarily relevant. Third, it demonstrated, as in the Ashkenazi case, the importance of ancestry: if African American women share genetic profiles with women in Africa, might that information be medically useful? And finally, it offered an explanation for differential cancer rates and types of diagnoses among some groups of black US women, especially those stricken with cancer at an early age, statistics that continue to baffle epidemiologists. If studies showed a difference in incidence, mortality, and diagnosis, even when confounding social variables like income were taken into account, might not ancestry help provide a missing piece of the epidemiological puzzle? Could genomics aid in the reduction of health disparities, a high priority research and policy area of federal institutions like the National Institutes of Health, and the National Cancer Institute in particular?

The Rhetorical Problem of Race

Generally speaking, geneticists understand "race" to be a social category, but one that nevertheless includes individuals for whom ancestry plays a role in health outcomes. For example, "African American" includes people who can trace their ancestry to West Africa. Heredity research thus identifies genetic markers that link individuals from different racial and ethnic groups from around the globe. The putative goal of BRCA research on people of different ethnic and racial identities is to contribute to medical care tailored to the individual, for whom any number of ancestral markers play a role in disease.

Although social categories may be an unavoidable necessity for recruiting study subjects, genomic methodologies are not employed, theoretically, to study race as such. They study populations. Yet BRCA researchers claim to be addressing the problem of health disparities, an epidemiologic phenomenon, not a genomic one. In the 1997 study by Gao's team, for instance, the researchers made explicit reference to breast cancer statistics that document disparities between white and black women:

> Among women born and raised in the United States, African American women have a lower risk of breast cancer than Caucasian women. However, the incidence rate of breast cancer in African American women is

rising among women <40 years of age. It has been observed that the age distribution of disease onset as well as tumor histology is different between Caucasians and African Americans. African Americans have a greater incidence at ages 30–44 years, and medullary carcinoma histology is more frequent in these patients. Furthermore, breast cancer in younger African American women may be more aggressive, leading to a decrease in the overall survival rates of African American women compared with that of Caucasian women.[19]

To make the case for a hereditarian explanation of these disparities, Gao and colleagues hypothesized that early age at diagnosis was a marker of hereditary predisposition. A diagnosis at a young age (defined differently by different researchers, anywhere from under thirty to under forty-five or, simply, as pre-menopause) suggests that genetic susceptibility has been a factor in that diagnosis. They also proffered the theory that since "the vast majority of African Americans originated from western Africa, where breast cancer is considered to be a rare aggressive disease predominantly affecting young women,"[20] genetic alleles probably accounted for the particular phenotypic distribution of breast cancer in the United States. Thus, a second major reason genomics would be relevant for health disparities researchers, in addition to the parallels seen between young African American breast cancer patients and already studied women with BRCA-related breast cancer, are the parallels observed between African American and African women. It would seem that shared phenotypes, along with the history of the African diaspora, imply shared, medically significant genotypes.

The claims by Gao's team and other researchers are fraught with logical and evidentiary problems, which this chapter will address in due course. The phenomenon whereby race as a biological concept displaces race as social formation cannot, however, be fully explained through recourse to the tools of argument analysis. Nor can it be fully explained through recourse to sociological methods. Rather, it must also be explored as a rhetorical and ideological problem wherein a persistent racialism in epidemiology finds material evidence in the burgeoning field of race-based genomics research. Indeed, in order to understand genomics as a racializing discourse, we must first examine the ways in which epidemiology has itself engaged in racializing practices, thus opening up conceptual and discursive space for genomics to do the same.

Epidemiologists generally employ racial identifiers in order to understand the relationship between lived experience of race, broadly construed, and particular disease outcomes. In breast cancer research, for instance, it is assumed that established risk factors for breast cancer are culturally variant, and thus women who identify themselves or are identified as black or African American are recruited as study subjects (I discuss the problem of the word "culture" and the study of race below in this chapter). Related to this, "health disparities"—a catch-all term[21] for observed differences in disease incidence and mortality rates between groups of people defined by race, ethnicity, gender, or socioeconomic status[22]—presumes that the lived experience of social identity accounts for these differences. Thus, the category African American is medically relevant, insofar as the lived experience of being black in the United States, whether or not one can trace one's ancestry to Africa, can influence susceptibility to diseases like cancer. That said, epidemiologists studying African American women and breast cancer have nonetheless allowed for—even suggested—genetic explanations for the disparities they observe.

The controversy over the 185delAG mutation of BRCA in Ashkenazi women is a case in point, as it opened the door to the subsequent geneticization of breast cancer in racial groups.[23] Before the reports of the 185delAG mutation, epidemiologists had claimed that being Jewish was a risk factor for breast cancer.[24] At the time, Jewish identity was discussed as a known or established risk factor (sometimes captured by the umbrella term "religion") by public service organizations like the American Cancer Society (why religion—in particular, Judaism—was thought to increase risk has never been sufficiently explained).[25] When, in 1995, geneticists found a link between Ashkenazi ethnicity and BRCA1 and 2 mutations, this risk factor found a biological explanation and, with that, newfound credibility. BRCA research thus provided a discursive site through which risk factors associated with the embodied experience of ethnicity were resignified as manifestations of inherited susceptibility. (In media reporting about the BRCA mutation data, it was at times assumed that the initial 1995 study by Struewing and coauthors had been undertaken with the intent of explaining putatively higher cancer rates among Jewish women).[26] At times, epidemiologists seemed as eager to embrace the hereditary explanation of the so-called Jewish risk factor as their geneticist counterparts were. For example, a research team of epidemiologists headed by Kathleen Egan of Harvard University studied Jewish women with a family history of breast cancer and women without such a history. The researchers concluded that risk for Jewish women with a family history of breast cancer was greater than for women in the control group and therefore "these results are consistent with data suggesting that certain groups of Jewish women have a higher than expected rate of mutation in the breast-cancer gene BRCA1." [27] Yet the researchers did not know whether the subjects in the study had actually inherited any BRCA mutations. In fact, no cancer researcher knew at the time whether women with these so-called Jewish BRCA mutations did have an associated increase in breast cancer risk (that study would be published a year later). [28]

At the same time that BRCA research was shifting its emphasis to populations, epidemiologists began taking the problem of black-white differences in breast cancer incidence and mortality much more seriously. In an important review essay on black-white disparities, the epidemiologist Bruce Trock suggested that breast cancer in black women is, materially speaking, largely a different disease. He claimed that when confounding variables such as income, education, socioeconomic status, and stage at diagnosis are controlled for, disparities persist. So, for example, even if black women have the same access to health care as white women, they still face more dire prognoses. [29]

These observations implicitly presume breast cancer risk in black women to be the materialization of heredity since genes are seemingly the only entities that can explain disparities once all social variables are controlled for. As Trock summarized the argument, the fact that black women "maintain a survival deficit compared to white women, even after controlling for stage of disease at diagnosis and other factors associated with socioeconomic status such as treatment and obesity, suggests the possibility of biological differences associated with race." [30] Breast cancer risk, for Trock, is a predisposition that exists prior to and is independent of the embodied experience of the otherwise social category of race. He describes an underlying biology at work, which is to say a cause or basis of some other thing, and concludes the essay with an endorsement of genomics research as a way to further understand these disparities. A 2006 review by a group of oncologists and epidemiologists similarly concluded:

> In several studies, racial differences in survival remained after adjustment for state at diagnosis, access to health care, treatment, comorbid illness, marital status, and other pathologic and sociodemographic variables. These data point to possible differences in the *nature of the disease itself*, supported by the observation that breast cancer in African-American women is more aggressive at presentation, characterized

by an increased prevalence of high-grade, hormone receptor-negative tumors[emphasis added].[31]

The use of the word "race" (or "racial") in this way may seem surprising given that epidemiology explicitly intends it to be a social category rather than a biological one. A close look at the history, politics, and methodological practices of the field, however, may help explain its implicit, unacknowledged racialism. The inclusion of race in epidemiological research (an aspect of what Steven Epstein calls the "inclusion paradigm") has a complex history, owing to several interlocking political and economic forces.[32] Epidemiology has not, however, successfully elucidated exactly how race plays a role in disease etiology. Janet Shim, for instance, has shown that accepted and routinized epidemiological methods (specifically, the multifactorial model of disease) require that the social reality of race be measured at the individual level. "This kind of devolution," she writes, "simplifies a complex world into smaller, presumably independent units of observation."[33] Researchers choose those variables "closest" to the observed outcome (disease), and these "typically translate to the direct biological risks or causes of disease and/or to the lifestyles or behaviours addressable at the individual level."[34] Race, in effect, becomes a black box, allowing epidemiologists the political luxury of claiming to explain health disparities even though their data do not shed light on what it actually means for race to explain health—it is, in practice, merely assumed. Indeed, "adjusting at the individual level for an effect that occurs causally at the society level cannot logically produce a meaningful model of disease etiology, no matter how refined the measures."[35]

What the multifactorial model does accomplish is the association of race with observable, isolatable, biological characteristics of the racially classified person, decontextualized from the array of factors associated with lived experience. For example, social isolation, a manifestation of structural racism, is one way in which race increases risk for breast cancer.[36] But to study the effects of social isolation, researchers measure glucocortisoid levels in individual women. These physiological changes in the body do not, in and of themselves, explain the complex connections between racism, social isolation, and one's embodied experience of these social relations. The examination of these bodily changes opens up the possibility—and a likely one as this knowledge makes its way into public health and medical practice—of correcting for elevated glucocorisoid levels at the individual level.[37] In turn, the complex processes by which race

structures embodied experience never become objects of study (and political intervention) in their own right. They remain, according to Shim,
structural constants, conceptually and materially outside the control of
both social and scientific actors.[38]

Merlin Chowkwanyun attributes the latent biologism of health disparities research to the dearth of historical perspectives on race as a social,
embodied identity category. He is worth quoting at length:

> We may learn about disease-by-disease, health outcome-by-health
> outcome, and year-by-year variations in racial health disparities (all of
> which remain vital data to continue collecting), but the fundamental
> findings themselves will soon not be very surprising. They consist of
> two elements: first, the simple (and durable) existence of the dispari-
> ties themselves, and second, the actual explanations given for them.
> The latter include poor health-care access, antagonistic racial attitudes
> of providers, stress caused by everyday discrimination and stereotyp-
> ing, proximity to environmental health hazards, neighborhood char-
> acteristics, and general socioeconomic status (SES)—all of which often
> vary starkly, the research indicates, by race. Methodologically, the work
> draws heavily, though not exclusively, on rich quantitative data sets
> generated from many sources, including public health surveillance, sta-
> tistical collection efforts, life-course studies, and surveys. What is miss-
> ing, however, is a deeper understanding of how and why these social
> determinants of racial health disparities matter so much, the long-term
> *process* through which they came into being, and how they might have
> been avoided. I argue, then, that the major shortcoming in racial health
> disparities research is an absence of historical perspective that would
> enable exploration of historically rooted "fundamental causes." This
> analytical lacuna, in turn, may become a major pitfall, hampering fuller
> understanding of causal dynamics at exactly the moment when interest
> in racial health disparities has reached unprecedented levels.[39]

Finally, Nancy Krieger—herself an epidemiologist but one who has
published a great deal on the field's methodological shortcomings, especially as they pertain to race[40]—attributes epidemiology's failure to provide meaningful answers to the question of racial health disparities to
its anti-theoretical stance, not unlike the critique of pragmatism's influence on medicine and public health that I discuss in chapter 1. Krieger
observes:

Shared observations of disparities in health . . . do not necessarily translate to common understandings of cause; it is for this reason theory is
key. Consider only centuries of debate in the US over the poor health
of black Americans. In the 1830s and 1840s, contrary schools of thought
ask: is it because blacks are intrinsically inferior to whites?—the majority view, or because they are enslaved?—as argued by Dr. James Mc-
Cune Smith (1811–1865) and Dr. James S Rock (1825–1866), two of the
country's first credentialed African American physicians. In contemporary parlance, the questions become: do the causes lie in bad genes?,
bad behaviours?, or accumulations of bad living and working conditions born of egregious social policies, past and present? The fundamental tension, then and now, is between theories that seek causes of
social inequalities in health in innate versus imposed, or individual versus societal, characteristics. Yet, despite the key role of theory, explicit
or implicit, in shaping what it is we see—or do not see, what we deem
knowable—or irrelevant, and what we consider feasible—or insoluble,
literature articulating the theoretical frameworks informing research
and debates in social epidemiology—and epidemiology broadly—is
sparse.[41]

Despite these serious, arguably crippling criticisms of health disparities
research and the epidemiological methods it typically employs, genomics'
incursion into this field of inquiry is not simply epidemiology's latent biologism taken to its logical conclusion. Epidemiology encompasses a diverse field of practitioners, and many of its critics, like Krieger, hail from
within the field.[42] The result is that the use of ethnic and racial categories
is routinely contested and debated, and they therefore remain conceptually in flux.

The so-called Jewish risk factor, for instance, was contested at the time
of the first BRCA report on the relevance of Ashkenazi ancestry.[43] Significantly, when that report was published, it seemed to provide evidence
for those eager to define Jewish men and women as a biological Other.
Public discourse about the relationship between BRCA and ancestry
reduced the complex interplay of embodied life and health to genes, a
metonymical substitution of part of an ethnic group (the Ashkenazi) for
the entire ethnic group (Jews). The 185delAG mutation was commonly
referred to in media reports as "the Jewish breast cancer gene" (a *Buffalo News* article went so far as to claim that it resides "solely" in Jews).
This episode demonstrates how an epidemiological risk factor, one that

presumably allows for greater consideration of context when thinking about risk, reemerges as evidence of biological difference between Jews and non-Jews.[44] Genomics research lends an air of authenticity to a race- or ethnicity-based risk factor for which there is little evidence of hereditary underpinnings but for which there is considerable scientific and popular appeal.

Not surprisingly, as the Ashkenazi research has developed, the so-called "Jewish" mutations have been found in non-Jewish groups—suggesting that carriers are all members of the same population. The 185delAG mutation in particular has been found in persons residing in Morocco, Spain,[45] and Iraq,[46] all of whom report no Ashkenazi ancestry. In one study of BRCA genes and ethnic and racial groups, researchers observed that the 185delAG mutation was detected in a significant number of study subjects classified as Hispanic.[47] Although one interpretation of these findings is that they merely reveal unknown Jewish ancestry, it is nevertheless also the case that the story of 185delAG has been a complicated and nuanced one, demonstrating the ways that the term "population" can serve to deconstruct anachronistic ideas of race and ethnicity.[48] What is more, even the concept "population" has shifted as this research unfolds, with recent studies casting doubt on 185delAG's status as a founder mutation.[49] As the next sections show, no such story has emerged in the discourse of BRCA susceptibility and African American women.

"African" Breast Cancer: Tumors as the New Materiality of Race

Inspired by the suggestion that founder mutations of the BRCA genes would enable the generalization of BRCA research across multiple population groups, and undeterred by the documented genetic heterogeneity of African American women, BRCA researchers pursued efforts to identify one or more exclusive mutations among this group of test subjects. Then, and only then, they reasoned, could race (as a proxy for ancestry) be a criterion, independent of family history, for offering a BRCA test. In 1999, after the publication of several reports suggesting the existence of such a mutation (labeled 943ins10), Heather Mefford and colleagues performed additional analyses to determine whether 943ins10 was in fact a founder mutation. They determined that it did display the characteristics of a founder mutation, in part due to a geographic distribution suggestive of patterns of migration associated with slavery and the African diaspora more generally.[50] Moreover, "the 943ins10 allele has not been observed in any patients with breast cancer who identify their ancestry as solely European."[51] The researchers argued:

> The migration patterns of African Americans and, hence, the current areas of residence of African American families, may explain the difference, among clinical centers, in the prevalence of the mutation. To determine, among African American women, the proportion of inherited breast or ovarian cancer attributable to BRCA1 943ins10, we would like to encourage testing for this mutation among African American breast and ovarian cancer[52] patients from various regions of the United States. Given the increasing incidence of and higher mortality from breast cancer among African American women, it would be useful to obtain as much information as possible about the roles of BRCA1 and BRCA2 in this population.[53]

The authors of this study were careful to employ the term "population"; indeed, founder mutations, by definition, are material artifacts of ancestry, not race. And ancestry, unlike race, is not a geographically bound concept. So, for example, the 943ins10 mutation is present in individuals who cannot trace their ancestry back to Africa, just as the 185delAG mutation has been found in people residing all over the world. Founder mutation research casts its testing net far and wide, identifying specific alleles of genes in people from around the globe in order to trace their evolutionary history. Thus emerge particular meanings of "population."

Founder mutation research has not, to date, resulted in the identification of mutations that would justify targeted screening of women for whom West African ancestry can be confirmed. Nevertheless, researchers have discovered a novel way of circumventing this apparent genetic dead end: the similarity of *breast cancer itself* in African American and West African women. Breast cancer, in other words, has become the observable phenotype of a suspected underlying genotype.

Two studies are representative of the belief that the morphological characteristics of breast tumors imply an ancestral connection between these two groups of women. In the first study, seventy young Nigerian women of African ancestry (all but four were diagnosed with breast cancer at or under the age of forty) were tested for BRCA1 and BRCA2 mutations.[54] The sample was ostensibly chosen because, they say, African women are more likely to be diagnosed with breast cancer at an earlier age and with deadlier types of tumors than are white women with no known African ancestry. This relative difference is observed both in Nigeria and the United States (where African American women are more likely than white women to be diagnosed with these sorts of tumors). The data thus suggest "shared genetic ancestry," despite the "significant genetic admixture" among blacks in the United States, and "could reveal genetic components of breast cancer common to all Blacks in the African diaspora."[55]

Similarly, the authors of a 2005 review essay in the journal *Cancer* titled "Breast Cancer in Sub-Saharan Africa: How Does It Relate to Breast Cancer in African-American Women?" note the ways in which breast cancer phenotypically links African and African American women, suggesting a population hypothesis despite the lack of any known genetic identifiers. The researchers write: "Despite the substantial heterogeneity of ethnic ancestry that exists within many African American families, a disproportionate risk for breast cancer mortality has been consistently demonstrated for women who identify themselves as African American."[56] They explain that mortality is associated with the types of tumors seen in African and African American women and that age at diagnosis further suggests an ancestral connection between these two groups of breast cancer patients. "Hereditary factors," they conclude, "may explain some ethnicity-related issues."[57]

The discursive movement from phenotype to genotype is possible because of the assumption that a *shared relative difference* justifies the hypothesis that health disparities are a genomic phenomenon: that the way in which black and white US women differ is the same way in which black and white women differ in West Africa—namely, in age distribution at diagnosis and in type of breast cancer. Black American women are more likely to be diagnosed at an early age and with aggressive tumors; black African women are more likely to be diagnosed at an early age and with difficult-to-treat tumors. So, the parallels drawn between African and African American women are not grounded in actual genetic markers (markers that are both common and exclusive to these groups of black women) but in these groups' relative differences with white women. These studies assume that materially similar breast cancers are the phenotype corresponding to an underlying genotype, allowing the possibility that African and African American women are a biologically definable population, even if the actual markers have yet to be identified. Thus, the failure of genomic methods to identify recurrent and/or founder mutations[58] no longer impedes asserting the value of BRCA research for African

American women. The conclusion of these authors, in fact, is that continued genetic testing of black women here and abroad is both medically and scientifically justified.

One consequence of presumed shared ancestry is that it geneticizes these breast cancers by drawing attention away from other reasons why these two groups of women suffer from similar tumors (and why, in many cases, they do not). Indeed, the very design of these studies is conceptually and methodologically limited insofar as comparisons of African and African American women overlook and/or ignore important differences that may exist within these arbitrarily defined groups. (Related to this is the problem of relying too much on study designs that compare black and white women, either here or in other countries.)[59] These BRCA studies tell us nothing about differences among black women—that is, how their risks vary due to the complexities of embodied experience.

Mapping the African heritage of breast cancer seems to hold more significance for BRCA researchers than ascertaining what, if any, medical relevance there is in doing so. Yet very real differences exist between black American women and black West African women. Breast cancer is much more common in the United States than in West Africa, and this itself raises a number of questions regarding the role of environmental factors in breast cancer trends. More important, the demographics of breast cancer, both here and in west Africa, are ever-changing, shifts that cannot be explained by evolutionary processes that take hundreds, if not thousands, of years to unfold. And not all black women, of course, are diagnosed with the type of breast cancer described in these studies. Nor are all white women excluded from this particular risk profile. One essay rightly concluded:

> Race and ancestry are confounded both by genetic heterogeneity within groups and by the widespread mixing of previously isolated populations. The assignment of a racial classification to an individual hides the biological information that is needed for intelligent therapeutic and diagnostic decisions. A person classified as "black" or "Hispanic" by social convention could have any mixture of ancestries, as defined by continent of origin. Confusing race and ancestry could be potentially devastating for medical practice.[60]

Here the case of sickle cell anemia and sickle cell trait is instructive: "If sickle-cell disease is suspected, then the correct diagnostic approach is not simply to determine the patient's race, but to ask whether they have African, Mediterranean or South Indian ancestry. To use genotype effectively in making diagnostic and therapeutic decisions, it is not race that is relevant, but both intra- and trans-continental contributions to a person's ancestry."[61] To be sure,

> sickle-cell anemia is not a "race specific" disease that reflects the underlying genetic make-up of West Africans and African Americans. The higher prevalence of sickle-cell disease in African Americans is in a very important sense an artifact of classification. In Brazil, for example, where many African Americans would be considered white the prevalence of a disease such as sickle cell might be very similar between people considered white. By virtue of ancestry, not all African Americans are at risk of sickle-cell disease, whereas some white populations and Latinos have a risk which is not well documented.[62]

"The false universalization of heritable traits," conclude Marcus Feldman, Richard Lewontin, and Mary-Claire King, "poses fewer health risks than a false particularization of such traits."[63]

Another consequence of these methodological choices is the reification of race. By grouping Africans and African Americans together in order to understand black-white differences in breast cancer rates, geneticists are, by default, privileging the African ancestry of black women. The reification of race is possible because of the ways in which scientific objects are rhetorically constituted. Rhetorical discourse is perspectival—the biological body is only partially accessed using the theoretical and conceptual tools available to us.[64] As a rhetorical term, "race" (operationalized in this breast cancer discourse as "African" and "African American") privileges certain characteristics of the materiality of human genetic diversity. Any classification system is, by necessity, only a partial snapshot of the complex patterns of human evolutionary history, rendering certain aspects of human diversity prominent (for instance, the primary continent of origin), others (such as admixture) less so. To be sure, geneticists can find and choose biological markers that will group individuals in any number of ways.[65] The rhetorical perspective of scientific discourse, moreover, helps us identify the inextricable connection between a scientific object (in this case, the racialized gene) and the epistemological and ideological investments of the scientific community.

Interestingly, the near-exclusive attention to African ancestry belies the lack of evidence of founder effects[66] on breast cancer risk in black women. More important, the medical relevance of ancestry depends, for obvious reasons, on its being a decisive factor in recommendations for prevention. Yet by some accounts, approximately 35 percent of African American women cannot trace their ancestry to West Africa. Complicating this further is the fact that "by some estimates more than three quarters of Americans who identify themselves as black or African American have a blood relative who identifies themselves as white or Caucasian."[67] But the literature on founder mutations often ignores the very real possibility that the "African" founder mutation 943ins10 could be detected in "white" women. According to one data set published in 2003, over sixty BRCA1 mutations had been detected in African American and African families as well as in families with no reported African ancestry. One of these mutations is the 943ins10 founder mutation of the BRCA1 gene.[68]

To better illustrate why the privileging of African ancestry is a unique and significant development, consider, for the purposes of comparison, the discourse of white breast cancer. According to the epidemiological data cited by BRCA researchers, compared to black women, white women in the United States are more likely to get breast cancer (except women under the age of thirty); are less likely to die from breast cancer; and are more likely to be diagnosed with less aggressive, more treatable tumors. Arguably, this epidemiological profile of breast cancer would suggest that ancestry might explain these breast cancers—in this case, European ancestry. But white women are not described in this manner. As of 2008, the word "white" or "Caucasian" has never appeared in a title of a BRCA research article, as there has been no attempt to explain "white" or "Caucasian" breast cancer as "white."[69] No studies exist that racially group white women in the United States and white women in Europe. Rather, we have studies on founder mutations linking women within or across particular nation-states.[70] (Whether or not one can really say there is a Finnish or Spanish BRCA mutation is beside the point; the attempt to do just that demonstrates that women can be grouped any number of ways. What the critic must do is consider how these choices are made and why they matter.)

The de facto assumption is that white women are a genetically diverse population formed from multiple ancestral lines, any of which may or may not play a role in breast cancer risk.[71] As a result, European ancestry was complicated early on with the identification of mutations linked to the Ashkenazim. For white women, then, significant family history, personal history, and Ashkenazi ancestry have remained the principal variables for justifying BRCA screening. White American women may not be as genetically diverse as other groups (again, this is ever-changing), but it is nevertheless not unusual to find novel mutations in white subjects (many of the reported mutations of BRCA1 and BRCA2 appear in just one family).[72] Yet the appearance of such mutations has not tempted geneticists to presume, however implicitly, that they are a race unto themselves. Methodologically, white women are more likely to be partitioned into several subcategories. A 2007 BRCA study of various ethnic and racial groups categorized the subjects as Hispanic, Asian American, African American, and non-Hispanic white with or without Ashkenazi Jewish ancestry.[73] This study, like those previously discussed, was intended to correct for the overrepresentation of women of European descent in the Breast Cancer Information Core database. Presumably, Hispanic is the ethnic category involved here (more on that below). Yet it is a category that also serves to deracialize the category "white." A less genetically diverse population than African Americans, white women are subcategorized in three different ways by the study's authors: Hispanic, non-Hispanic with Ashkenazi ancestry, and non-Hispanic without Ashkenazi ancestry. In another study, published in 2009, white subjects were divided into five groups.[74] In neither case was the far more diverse group of African Americans afforded such nuanced treatment. Instead, the familiar language of inherent biological differences, again drawn from epidemiological data, was employed.[75] African Americans, these reports affirmed, are biologically distinct given their propensity for certain types of tumors at particular ages. Age alone, it has been suggested, can serve as a criterion for offering the BRCA test to black breast cancer patients[76]—a criterion for screening that rhetorically racializes this group of women.

Another way in which black women are indirectly racialized is by the collection of data on the 185delAG mutation. In BRCA studies of racial and ethnic groups, this Ashkenazi-related BRCA1 allele both complicates the evolutionary history of otherwise "white" persons while simultaneously simplifying it for African Americans. For example, in the studies by Esther John and coauthors and Michael Hall and coauthors, 185delAG was detected in all groups except African Americans. This was true even for the subjects in the ostensibly racial groups "Asian" and "Native American." Hence, these other groups may be understood to be populations, not races or ethnicities, by virtue of their ancient Ashkenazi heritage. Another group garnering significant attention in the BRCA literature is people of

Hispanic background. At least one study has found a suspected founder mutation in people reporting Hispanic ancestry. In a caveat, the authors wrote that although "Latino" is a term employed the US Census and is more of an "ethnic" or "cultural" designation, they purposefully employed "Hispanic" as a way to understand the intersection of this social group's particular evolutionary history and breast cancer risk.[77] Although this may very well have the effect, at some point, of homogenizing, or racializing the identity "Hispanic," so far no one part of this group's ancestral history has been privileged over another. Instead, geneticists have engaged in a complex, nuanced examination of the evolutionary history and dispersed geographical locations of their Hispanic study subjects. One such example is the use of the term "mestizo." Here, researchers are acknowledging the existence of multiple ancestries and an attendant genetic admixture rather than privileging continent of origin, as they do in the case of African American women. The suspected BRCA1 founder mutation del (ex 9-12) that is prevalent among study subjects identifying themselves as Hispanic has also, as it turns out, been reported in persons who identify themselves as African American and Native American. The mutation's investigators conclude that because BRCA mutations may account for a "higher proportion of the breast cancers in young Hispanic women," this group is "analogous" to the Ashkenazim.[78] Notably, they do not invoke young African American women for comparison purposes.

What I have described is just one way in which BRCA research racializes breast cancer. In the next section, I describe another, closely related discursive development. I argue that in the absence of founder mutations, it has not been scientifically possible to group women together by way of specific mutations they all share. Black women in the United States and in West Africa are not, as it turns out, linked by specific, shared genetic markers; they are linked by virtue of their genetic diversity—in this case, the diversity of BRCA variants identified by mutation databases. The result is the racial demarcation of women based on relative genetic diversity.

"Diversity" as Racial Signifier

Before exploring how genetic diversity acts as a racial signifier, it is first necessary to understand its origins in debates about race and science, in particular how it has grounded some of the most powerful and effective scientific challenges to the race concept. Beginning with naturalists in the eighteenth century, race was principally used to make sense of observed differences in skin color and other outward physical characteristics, which were taken to be markers of underlying biological difference among continent-based populations. This particular notion of race has been the subject of heated debate within various scientific communities, beginning with critiques of early twentieth-century eugenics, prominent attacks in the 1950s and 1960s by anthropologists such as the late Ashley Montagu,[79] and consensus statements by professional societies and the United Nations Educational, Scientific, and Cultural Organization, all resulting in public claims that race is not a legitimate biological concept.[80] These critiques are grounded in the observation that science has been unable to find biological markers that are found in all persons of one race (and, conversely, cannot be found in any members of any other races).

There are two notable ways in which genetic diversity has been marshaled to reject the notion of race. First is Richard Lewontin's well-known claim that most genetic diversity (85 percent) is found within traditional race groupings, not across them.[81] Second is the publicity surrounding the HGP's revelation that humans are, genetically speaking, 99.9 percent the same. On one hand, then, humans do not diverge much from one another in genetic terms. On the other hand, most of the gene variation that we do see is found within socially defined racial groupings.

Although these pronouncements have been socially and politically significant, they mask the epistemological and financial interests in diversity. Lewontin's argument has never foreclosed the possibility that diversity may be important for understanding the relationship between ancestry and disease. And the HGP was criticized early on for not, in and of itself, telling the story of genetic variation.[82] Both the appearance and the preservation of genetic variations in geographically and culturally isolated populations interest medical researchers who want to understand the genetic basis of disease and the relationship between ancestry and risk (for example, the presence of the sickle cell trait in individuals of African descent). Historically, the concern has been with rare but highly penetrant[83] genetic mutations, such as those associated with the so-called Mendelian disorders. The development of sophisticated sequencing technology has subsequently enabled researchers to turn their attention to low-penetrant but more common alleles that play some role in chronic diseases such as heart disease and cancer.[84]

The development of technology with which to study the complexities of human diversity has not resolved debates about race, however. It has, on the contrary, renewed—with much fervor—debates about whether or not patterns of human genetic diversity are compatible with categories of race, the latter defined as the primary continent of origin.[85] There is a growing debate as to whether or not a genetically inferred primary continent of origin demonstrates a material reality to socially constructed racial categories—what Jonathan Marks, a critic of the use of race, has termed the "gene pool residual" understanding of race.[86] For many researchers, the answer is yes, which has raised the further question of whether the label "African American" can be a proxy or surrogate for African ancestry, a question reaching lay audiences with the March 2005 publication of an opinion piece by the biologist Armand Leroi in the *New York Times*.[87]

Leroi claimed that race is in fact a useful proxy or shorthand for human genetic diversity, dismissed the methodology behind Lewontin's 1972 essay as outdated, and cited several articles demonstrating a near correspondence between racial identity and genetic cluster groupings. Cluster analysis, in which researchers sort study subjects into genetically similar subgroupings, can be performed with or without knowledge of individual racial or ethnic identity. The fact that it can be done without such knowledge has led some researchers to conclude that racial identity should never be a part of the geneticist's research methodology.[88] For others, including the geneticist Neil Risch, it is too soon to jettison the use of racial and ethnic labels. Although both sides of the debate share the goal of individualized, whole-genome analysis, in the short term, these researchers suggest, proxies are both technologically and economically necessary.[89] Moreover, they argue, a colorblind method would impede researchers' attempts to improve minority health. Failure to acknowledge the race or ethnicity of research subjects makes it difficult to take into consideration nongenetic variables connected with lived experience—variables that can affect health outcomes. According to this reasoning, geneticists have an obligation to study race in both its social and biological dimensions:

> Studies using genetic clusters instead of racial/ethnic labels are likely to simply reproduce racial/ethnic differences which may or may not be genetic. On the other hand, in the absence of racial/ethnic information, it is tempting to attribute any observed difference between derived genetic clusters to a genetic etiology. Therefore, researchers performing studies without racial/ethnic labels should be wary of characterizing difference between genetically defined clusters as genetic in origin, since social, cultural, economic, behavioral, and other environmental factors may result in extreme confounding.[90]

In other words, since it is inevitable that genetic clusters emerging from DNA sequencing will correlate with self-identified race and ethnic identity (or very nearly so; there are always exceptions), researchers should retain racial identifiers in order to parse out the relative contributions of heredity and environment to observed disease outcomes.

On closer examination, the sophisticated cluster analysis techniques invoked by geneticists—techniques that putatively reveal the existence of race and thus compel us to revisit old debates—have been shown instead to function as a technology for the materialization of racial norms. According to Deborah Bolnick's critical account, several of the most commonly cited studies (for example, Bamshad et al. 2003) do not support the correlation between genetic cluster and race that has fueled the much publicized and titillating interpretations of Leroi, Risch, and others.[91] As it turns out, the software employed by some of these studies (called *structure*) sorts DNA sequence information into a predetermined number of groups. In some cases, the same set of DNA samples could have been, theoretically, grouped into one, four, or even nineteen clusters. More important, and this bears on the question of medical relevance, even if it is possible to "find DNA sequences that differ sufficiently between populations to allow correct assignment of major geographical origin with high probability," it is nevertheless also the case that "genes that are geographically distinctive in their frequencies are not typical of the human genome in general."[92]

The use of more sophisticated software has not necessarily solved the problem of racialization. As recent research by Joan Fujimura and Ramya Rajagopalan has shown, "although the invention of a way of conducting human genetics research without using race categories provides an *opportunity*, its potential for transforming genetics research may be limited by downstream translation practices, the ways some researchers have subsequently used the Eigensoft software, the tenacity of concepts of race, and the many historical examples of seemingly neutral terms eventually coming to gain evaluative meanings."[93] This, despite what Fujimura and Rajagopalan describe as the best intentions of the geneticists employing the software.

Other genetic diversity research includes the HapMap project, an international research effort spanning six countries that has produced a map of haplotypes—sets of single nucleotide polymorphisms (SNPs)—thought to be closely linked to functional sections of the genome that can serve as signposts for the presence of suspected disease alleles. The second phase of the project, documenting over a million SNPs thought to be common in humans, will allow researchers to more easily pursue case control and exploratory studies in order to both determine the degree of risk associated with suspected pathological alleles and to locate more regions linked to disease onset.[94] The guiding principle of the HapMap project is the biomedical significance of diversity, yet the sampling methodology suggests that diversity is important only insofar as it marks off differences among the continents. Thus the HapMap consortium, while recognizing that no one population would be necessary for inclusion in order to map common SNPs, sampled a number of populations that nevertheless correlate with common understandings of race.[95] Like the cluster studies I describe above, the HapMap project demonstrates the essential paradox of genetic diversity in a culture structured by racialism: diversity, more often than not, serves to reify group boundaries, not break them down.

A close examination of BRCA discourse helps elucidate how this reification occurs. It was perhaps inevitable that diversity would be a topic of interest for BRCA investigators in the wake of the population turn. As the research progressed and more and more study subjects were tested (although recruitment has remained quite low), it became clear that not only did a statistically significant number of black women test positive for BRCA mutations, but the tests produced much different data than those found in other populations. In one study, for example, nineteen BRCA variations were detected in forty-three black families, as opposed to just nine variations in seventy-eight white families.[96] Many of these genetic variants were not reported in other groups. (This is why African American women are more likely to test positive for variations for which no clinical research is available.)

The range of diverse BRCA alleles—as well as their novelty—has not disproved the anachronistic meaning of race so much as provided new material evidence for it. As the data about BRCA diversity have accumulated, the language of the unique and distinct genetic profile has proliferated. Consider the following excerpts:

> Our data and those of other groups support the presence of a diverse spectrum of BRCA1 and BRCA2 mutations and sequence variations unique to Blacks of African descent.[97]

> The spectrum of mutations we observed in African Americans is vastly different from what we observed in individuals of European descent.[98]

Our data, coupled with that of other investigators, reveal a large num-
ber of distinct pathogenic mutations and variations among African
Americans. Most of these variations have not been reported among
Caucasians.[99]

In all three passages, diversity is a phenomenon that, paradoxically,
suggests both the impossibility and the necessity of defining persons racially. A "diverse spectrum" defies racial markers yet is "unique" to the
racial group in question; a "spectrum" implies diversity yet is simultaneously "vastly different" from white subjects. "Distinct" is a particularly
interesting word that is used in these studies, for it means, among other
things, "recognizably different in nature from something else of a similar type." If a genetic variant is "distinct" from other genetic variants, it
is different in nature. However, these BRCA variants are not merely self-
referential; they are parts of actual bodies. If they are distinct, then so too
are the bodies of which they are part. "Distinct," is also defined as "readily
distinguishable by the senses" and "so clearly apparent as to be unmistakable."[100] This is the very language used to describe race colloquially. By
employing the term "distinct" in scientific discourse, researchers convey,
however implicitly, that they, too, "sense" an "unmistakable" race when
they analyze gene frequency data.

Diversity, then, becomes material evidence of population migrations
and marks a boundary between African Americans and so-called Caucasians. In other words, diversity is not incompatible with the notion of racial distinction: although scientists may not be able to point to a particular
allele that all members of a "race" share, they can point to alleles reported
in one group and not another (in this case, alleles of the BRCA genes)
as evidence that racial differences in biology can explain biomedical phenomena.[101] So, African American women display a "unique" spectrum of
BRCA mutations, and this suggests that although there is substantial variation within groups of women identified as African American, this variation is distinct from patterns observed in other racial and ethnic groups,
both in terms of the structure of variations (the particular nucleic acids
that characterize them) and in the amount of variation (the number of
different or "distinct" BRCA alleles detected).[102] Here, then, is the inherent limitation of Lewontin's analysis: he argued that research consistently
demonstrated that groups defined by race (African, European, Asian, Native American) differed more from each other than they did from members of other races. Diversity was used, by Lewontin and others, to refute

the existence of race on the grounds that no set of unique genetic markers could be found to group persons together, unequivocally, under the umbrella of race. But although there is greater diversity within races than between them, it is still the case that the greatest degree of diversity can be found among individuals of African descent. Follow-up studies to Lewontin, for instance, have shown that the most diversity is found in Africa and African Americans due to genetic admixture and the intersection of multiple ancestral lines.[103] So greater diversity signifies African race, and lesser diversity characterizes other races, at least according to those scientists motivated to find ways to prove both the existence and medical relevance of diversity.

In practice, BRCA alleles labeled "unique" to one group of people have subsequently been reported in other groups. In a study of BRCA mutations in German subjects, for instance, a mutation detected in a "white" subject happened to be the same one that Gao and coauthors had previously labeled "unique" to blacks. The term "unique," then, does not really mean exclusive. Even if "unique" in this context means "common to," relative diversity, as I have shown, nevertheless distinguishes blacks and whites and thus constitutes the former as a race. Future studies of African Americans are likely to reveal many more BRCA variations than will be found in whites. Gene frequency research, along with research that privileges African ancestry by drawing parallels between African and African American women, is unlikely to yield nuanced studies of African Americans as an ethnic population for which African ancestry is just one part of a complicated evolutionary history that may or may not be medically important. These genetic discoveries are more than just evidence of the vagaries of genomics research; in a larger social context in which scientists and the lay public alike still believe in a biological basis for race, the use of words like "unique" and "distinct" matter, both in terms of validating popular beliefs and of discursively constituting racial subjects anew. "Admittedly, racial categorization is not the same as racism," says Jacqueline Stevens, "but racial categorization never happens without also producing racial hierarchies."[104]

A significant percentage of African American women can trace their ancestral roots to West Africa; they thus constitute a "population" in the same way that women of Ashkenazi descent do. However, lay understandings of population often displace more specialized ones in popular and scientific discourse alike. Thus, the mutation 185delAG came to be known as the "Jewish" mutation even though it was detected in a variety of individuals all over the globe, some of whom did not identify themselves as Jewish. For African American women, the situation is complicated further by genetic admixture. It is not surprising that, as of this writing, only one mutation has been identified as a possible candidate for explaining founder effects in black women, defined as such because it was detected in unrelated black women living in the United States, Jamaica, and the Ivory Coast (although not detected often enough to justify targeted screening). Also not surprising is the fact that it has been detected in white women claiming to have no African ancestry whatsoever.

It may be the case that these "white" women are mistaken about their ancestry. Nevertheless, the story of this founder mutation raises crucial questions as to the medical relevance of racialized BRCA research. I use the word "racialized" intentionally, both to reflect the actual terminology of researchers (the word "race" is used in genetics as well as epidemiological research about the putative differences between white and black tumors) and to make the argument that scientific discourse produces racialized subjects. Whether African American women are described in this research literature as an ethnic group, a population, or a culture, what emerges is a racial subject in the biological sense: a group of people linked to a continent and bearing certain genetic signifiers that distinguish them from members of other racial groups, even when the evidence provided is relative statistical frequencies and not a notion of a clearly delimited, "pure" race.[105]

Progressive Epidemiology and the Historicization and Politicization of Race

Breast cancer patterns are one way in which women's bodies reveal the persistence and destructiveness of institutional racism. Since 1998 the "Annual Report to the Nation on Cancer"—a collaboration of the National Cancer Institute, the Centers for Disease Control and Prevention, the American Cancer Society, and the North American Association of Central Cancer Registries—has documented the undue burden of breast cancer in the United States for women of color. In 1998 deaths from breast cancer were 28 percent higher for black women than for white women.[106] In fact, African American women have the highest breast cancer mortality rate of any racial or ethnic group in the United States.[107] Moreover, although blacks' overall rates for breast cancer are lower than whites' overall rates, breast cancer rates among young black women are higher than among young white women, and have been so for some time. As Krieger has observed, "combine relatively high incidence and relatively high mortality, and the net result is that US Black women have among the highest breast cancer mortality rates in the world."[108]

Viewed historically, breast cancer mortality rates among black women in the United States evince a disturbing pattern. Between 1973 and 1990, incidence in black women under fifty increased by over 16 percent, while the increase for white women in this age group increased by 6 percent.[109] Between 1969 and 1997, the age-adjusted death rate for white women dropped 15 percent, while it increased by 22 percent for African American women.[110] Rates are very high among older black women. Gross rates of breast cancer in women younger than fifty between 1992 and 1997 were actually lower for black than for white women, although adjusted for age they were slightly higher. These trends defy a genetic explanation: "beyond the clinical trials, epidemiologic assessment of the US population suggests biologic/genetic differences may have been overemphasized. Often unappreciated is the fact that black-white mortality was very similar in the 1970s and the disparity in mortality rate has grown every year from 1981 onward."[111]

The cancer community rallied around these rates, evidenced in part by a major conference in 2003 devoted to the subject of health disparities and culminating in a special issue of the *Journal of the National Cancer Institute*. "Race" is an operative word in this discourse, and deservedly so, as women who identify themselves or are identified by others as black or African American arguably experience life events differently than women classified otherwise. Race is a determining factor, for instance, in relative experiences of poverty and discrimination, which in turn can affect nutrition, stress, access to preventive care, access to palliative care, and overall well-being—all essential to whether a woman develops disease and, if she does, can survive it.

This historical trajectory is material evidence of the social dimensions of black-white breast cancer rates. The field of health disparities research[112] emerged as a direct response to the persistent, often startling statistics charting an ever-widening gap between blacks and whites in breast cancer incidence and mortality. As I argued above, epidemiologists and geneticists alike have questioned whether epidemiology possesses the methodological tools necessary to fully explain differences in breast cancer rates. Yet the claim that genomics must step in to explain these differences implicitly assumes that epidemiology is, paradoxically, both methodologically crude and sophisticated. On one hand, it is ill equipped because of its reliance on social variables like income—variables that, even when properly measured and controlled for, fall short of an adequate explanation of persistent disparities. On the other hand, epidemiology, it would seem, has achieved such a level of methodological sophistication and nuance that it can adequately measure and account for all the possible ways that social variables might explain racial differences in breast cancer incidence and mortality. This belief, implicitly expressed by epidemiologists and geneticists alike, rests on the rather improbable assumption that variables like education and class operate in ways that are generalizable across different racial and ethnic groups.

There are, however, many reasons why racial groups are not interchangeable when variables such as education, socioeconomic status, and income are measured. Similar levels of education do not necessarily result in similar incomes and employment opportunities, and higher debt-to-income ratios may mean that similar incomes carry less benefit for black Americans. Also, when one adjusts for stage of disease and access to health care, racism and class inequality can, theoretically, explain the kinds of breast tumors diagnosed in black women—what researchers call the problem of "residual confounding":

> Race is a social variable, not an inherent biologic variable. It is clear that compared with white women, a disproportionately higher proportion of black women present with later-stage disease and that within any one stage a higher proportion of blacks than whites have poor prognostic markers such as higher pathologic grade and estrogen receptor-negative disease. [Researchers] have shown that low education level and low income are associated with the diagnosis of estrogen receptor-negative tumors, which suggests that factors associated with socioeconomic status may influence the biologic behavior of breast cancer.[113]

Not only is it difficult to control for all sociocultural factors that can contribute to disease risk, the quotation above from the cancer researcher Otis Brawley supports the argument that even when researchers are reasonably confident that they have controlled for them, class position and racial identity nonetheless operate as confounding variables in ways that they sometimes ignore or simply do not yet understand. For example, the Institute of Medicine concluded in 2001 that differences in income and access to material resources cannot fully explain disparities:

Perhaps the most striking finding that emerges from the analyses of social environmental influences is the graded and continuous nature of the association between income and mortality, with differences persisting well into the middle-class range on incomes. The fact that socioeconomic differences in health are not confined to segments of the population that are materially deprived in the conventional sense suggests strongly that socioeconomic differences are not simply a function of absolute poverty. Moreover, because causes of death that are purportedly not amenable to medical care show socioeconomic gradients similar to those of potentially treatable causes, differential access to health care programs and services cannot be solely responsible for these differentials in health. Finally, because the gradient in morbidity and mortality persists even between middle class and well-to-do men and women, and even in societies in which material conditions are very good, it seems unlikely that gradients are due solely to material circumstances per se.[114]

"Race" confounds epidemiological data, not because underlying genetic/biologic variables are at work, but because "race" is not yet adequately theorized. Specifically, researchers must think creatively about how racial discrimination structures the everyday life of African Americans of all socioeconomic status designations—what feminist studies scholars would say is the need to understand the intersection of oppressions in the context of health. The study of racial differences in health must consider their effect beyond presumed access to resources and colorblind assessments of socioeconomic status.

Lewontin astutely observed that "asbestos and cotton lint fibers are not the causes of cancer. They are the agents of social causes, of social formations that determine the nature of our productive and consumptive lives, and in the end it is only through changes in those social forces that we can get to the root of problems of health. The transfer of causal power from social relations into inanimate agents that then seem to have a power and life of their own is one of the major mystifications of science and its ideologies." Shifting the focus to lived experience can return causal power to the social sphere. One model for studying race as lived experience as the outcome of embodiment, not heredity, is the "ecosocial" model put forward by Krieger. Focused on the guiding question of "who and what drives current and changing patterns of social inequalities in health," ecosocial theories combine the concerns and methods of political economy approaches (espoused by Lewontin and Richard Levins) with what she describes as "rich biological and ecological analysis."[115]

The ecosocial model requires the researcher to consider disease as the outcome of embodied life, or how we "literally incorporate, biologically, the material and social world in which we live, from conception to death."[116] As a model for studying embodiment, ecosocial theory considers such things as evolutionary history and ecological context; the distribution of resources and patterns of production and consumption; embodiment expressed at various levels, such as the local and national, and in different domains, such as home and school; and, finally, the ethical dimensions of knowledge production—how and why institutions and individual scientists, including epidemiologists, need to claim responsibility for the benefits and limitations of their theories, models, and methods. The ecosocial model thus brings together an otherwise disparate group of disciplines to investigate and explain health disparities, a model that opens up for consideration a much broader field of interventions. Although it is important to understand the biological mechanisms involved in disease causation (Krieger rightly notes that reducing economic inequity does not entirely explain improvements in the public's health over time; we must consider better sanitation, for example), it is also the case that "no aspect of our biology can be understood absent knowledge of history and individual and societal ways of living."[117]

The phenomenon of higher rates of hypertension among African Americans shows the benefit of ecosocial perspectives.[118] The scientifically discredited yet still popular evolutionary explanation is that African Americans with hypertension do not metabolize salt properly due to selection pressures from the murderous slave trade trips from West Africa. In contrast, the ecosocial explanation considers how hypertension rates are the outcome of both social and economic deprivation (diets lacking in fresh foods can increase risk), toxic substances and hazardous conditions (exposure to lead can cause renal disease), "socially inflicted trauma" (the body's response to racial discrimination and violence or the threat of it), inadequate health care, and, finally, whether opportunities for collective action in response to these inequities (access to community resources and involvement in social movements) can reduce risk for hypertension.[119]

The ecosocial explanation of breast cancer opens up a similarly rich and nuanced explanatory and actionable landscape, and in doing so casts the limitations of both traditional epidemiology and genomics into sharp relief. An interdisciplinary group of researchers in Chicago has proffered the theory that breast cancer mortality among black women is inextricably linked to social and economic environments. Their work, situated at the Center for Interdisciplinary Health Disparities Research, is part of a consortium of mostly publicly funded university-based centers, collectively called the Centers for Population Health and Health Disparities (CPHHD).[120] In 2007, the consortium published a progress report, titled "Cells to Society: Overcoming Health Disparities," in which they wrote:

> The mapping of the human genome provided an important platform for addressing the causes of cancer disparities in individuals and populations. New personalized medicine technologies also promise to help many individuals with cancer. However, while such approaches are essential, they are insufficient on their own for the development of an effective and efficient long-term strategy for gaining knowledge and preventing disease at the population level. Indeed, the explanations for disease occurrence in a population may be quite different from the causes of interindividual differences in disease occurrence.[121]

CPHHD research is "vertically oriented, starts at the top with race, poverty, disruption, and neighborhood crime; moves to isolation, acquired vigilance, and depression; then to stress-hormone dynamics; and finally to cell survival and tumor development."[122] The consortium's researchers "found that social isolation, vigilance, and depression were highly correlated and represent facets of a 'psychosocial suite' that represents how the social environment gets 'under the skin' to alter the body's ability to repair cells."[123] One hypothesis in particular that has captured their attention is the role of glucocortisoids in race-specific stress and the connection of that role to cancer development. Activation of glucocorticoid receptors sets off a chain reaction of biological responses that ultimately helps cancer cells survive. Thus, "higher reactivity to stress may predict earlier tumor development through heightened secretion of glucocortisoids."[124]

Another ecosocial hypothesis is that black women are more likely to be exposed to a certain class of chemicals known as organochlorines (OCCs), which increase breast cancer risk. The theory is that OCCs, such as the pesticide DDT and polychlorinated biphenyls (PCBs, now-banned chemicals used in the manufacture of electronics), are estrogenic and thus increase breast cancer risk (the effect being similar to reproductive history, which influences estrogen exposure).[125] Research has shown that the body burden of carcinogens is greater for black women than for white women, sometimes quite significantly so.[126] In one 2003 article, the authors observe that "studies over the past 30 years consistently have found OCC compounds to be present at higher levels in African Americans compared to whites, and this pattern appears to continue."[127] Robert Millikan and coauthors showed that levels of DDE (the principal metabolite of DDT) and PCBs were higher in African American than in white women and that this could explain the former group's heightened risk for breast cancer.[128] These dangerous chemicals may also explain higher rates among young African American women, since developing breasts are more vulnerable to environmental toxins.[129] Collectively, "these findings imply that breast cancer in women occurs as a generalized response to systemic carcinogens operating within the breasts' most active tissue."[130]

Genomics as a Racial Project

If breast cancer is partly the biological manifestation of the everyday experience of racism—how racial oppression "gets under the skin"—then antiracism, understood in the broadest sense of the term, could be tantamount to breast cancer prevention. Just what would this kind of prevention look like? It would include, among other things, prohibiting the commercial and industrial use of chemicals until they are proven safe; stopping the concentration of dirty industries in poor communities and communities of color; and implementing economic policy initiatives that produce substantial and far-reaching redistribution of resources (affirmative action; living-wage laws; affordable, environmentally safe housing; stronger labor laws, public works initiatives; and effective laws against racial profiling of any kind, to name just some). The abdication of state responsibility to eliminate the devastating effects of racism is analogous to what Steven Martinot has described in the legal context as the "decriminalization" of the state.[131]

In a period of financial crisis, and as attacks on social programs, economic justice, and regulatory infrastructure continue, it is reasonable to assume that these sorts of policies will remain politically and socially unfeasible. What I would like to raise here is the question of how scientific knowledge grounds—if only implicitly—arguments about resource distribution in an age of austerity (with the exception of the military-industrial complex duly noted). For this I turn to Michael Omi and Howard Winant's notion of a racial project:

[A racial project] is simultaneously an interpretation, representation, or explanation of racial dynamics, and an effort to reorganize and re-distribute resources along particular racial lines. Racial projects con-nect what race means in a particular discursive practice and the ways in which both social structures and everyday experiences are racially organized, based upon that meaning.[132]

Constructions of race are crucial to the successful operation of a racial project insofar as they undergird institutional practices and policies. In the 1950s, for example, unwed black mothers were blamed for soaking up precious economic resources and exacerbating the perceived economic insecurity of whites. In this way, argues Rickie Solinger, reproductive ca-pacity justified both status quo economic arrangements as well as policies that would concentrate wealth even more among white people. Crucial to this discourse was the belief that these black mothers were simply fulfill-ing a biological imperative bestowed on them at birth. Black female sexu-ality, implicitly taken to be an inborn trait, pervaded political discourse at the time. Even when "culture" replaced "biology" during brief periods of progressive social form, argues Solinger, it was really just a "dignified cover for biological racism."[133]

In this chapter I have discussed what I call progressive epidemiology's willingness and ability to reflectively consider how science inadvertently reifies race. This work has made it possible to show that structural and institutional racism results in extreme suffering. Indeed, it kills. Yet health disparities scholars—who hail from multiple scientific disciplines, and many of whom are ardent critics of the race concept—remain deeply in-vested in biological studies of health disparities. It is, of course, reasonable to expect them to be so. Ironically, insofar as these epidemiologists com-pel intervention at the social level, they may undermine the necessity of their work over the long term. Elevated glucocortisoid and OCC levels are not the causes of breast cancer—they are, to borrow Lewontin's phrase, agents of social causes. To intervene at the level of social causes, then, is to move beyond epidemiology's focus on illuminating proximate causes and into the arena of social movement and policy change. Research is surely required to substantiate the importance of such interventions. But that research may need to come from the humanities as well as the so-cial sciences (in the interest of full disclosure, I consider myself a scholar working in the humanities). Making the case for the relevance of histori-cal methodology for understanding health disparities, and specifically scholarly work that documents the material changes that determine what "race" means and how it is connected to public health, Chowkwanyun writes:

> If these rich works have one common takeaway, it is that racialism persists in American medical and public health history, and by extension, so does the "naturalization" of racial health disparities. By identifying the historically fluid and constantly changing nature of race as a category—and the meanings, stigmas, and prejudices bound up with it—these authors undercut the notion of race as a fixed entity across time and the pernicious idea that innate, immutable characteristics are linked to it.[134]

Stevens has called this the synchronic approach to race[135]—one that is most likely to flourish, it seems to me, in work that does not examine proximate causes, whether they be hormone levels or gene sequences. More important, a historical sensibility, says Chowkwanyun, creates the conditions of possibility for change: "By locating such factors and the human agents, decision-making, and the exercise of political power behind them, we are reminded that these disparities are not natural but created and thus undoable, *however awesome the task*" (emphasis added).[136] For Chowkwanyun, when health disparities research is fully situated within the critical historical tradition, it shows that disease outcomes are inextricably bound up with the geography, politics, and economics of segregation, redevelopment, housing, and employment. These, in turn, become sites of intervention.

Chowkwanyun thus makes the case for material interventions that—I would argue, following Barbara Fields[137]—are important means for interrupting the enactment of racial ideology. The analysis I have presented in this chapter shows that even when scientists intend their work to be antiracialist—for example, by avoiding the use of the word "race," employing alternative concepts such as "ancestry," and developing more sophisticated sequencing technology—they nevertheless effect the racialization of their research subjects. In her analysis of contemporary genomics research, for instance, Bolnick concluded that "ancestry is not that different from race in practice. . . . Since the contemporary Euro-American definition of race is based on continental geography, anthropologists and human geneticists use the term 'ancestry' much as the general public uses the term 'race.'"[138] Anne Fausto-Sterling, Jennifer Reardon, and Joan

Fujimura and Ramya Rajagopalan have also concluded that race seems to be so firmly entrenched in the mental habits of scientists that methodological improvements are insufficient in dismantling the latent belief in, and ultimate materialization of, race.[139] (In an interesting twist, the profit motive may serve as a bulwark against racialization. In a 2009 study sponsored by the company Myriad Genetics and published in *Cancer*, the authors concluded that all high-risk women are candidates for BRCA testing regardless of racial or ethnic identity.[140] Although as Epstein points out,[141] *pharmaceutical* companies will continue, for the foreseeable future, to create niche racialized markets.)

Proof of the persistence of race supports Stevens's proposal that the National Institutes of Health refuse to fund any genetics study that purports to explain health disparities unless "statistically significant disparities between groups exist and [the] description of these will yield clear benefits for public health."[142] This would prove a very high bar and would result in drastic reductions in current funding levels. Defunding genetics research is a different type of material change than, say, switching software from *structure* to EIGENSTRAT to determine gene frequencies of populations. It could effect, rather, a change in the external material conditions and social structures (much like the antiracism initiatives I listed above) that are the reason why racial ideology persists.

I have no doubt that that the scientists and race critics I have cited would disagree. The Chicago study on social isolation, for example, understood hormone levels to be products of gene expression; Krieger's ecosocial model is premised on the integration of social and biological research; and Fausto-Sterling, after a thorough treatment of the persistence of race and why genomics research diverts attention away from the link between racism and disease, nevertheless concludes: "We need, instead, to develop the habit of thinking about genes as part of gene-environment systems, operating within networks that produce new physiologies in response to social conditions."[143]

Indeed, what I have not yet considered is whether genomics permits us to consider "gene-environment" interactions, a dialectical relation that does not, in and of itself, privilege one part over the other and that has greater explanatory power than either genomics or epidemiology alone. Gene-environment interaction is embraced by a great many stakeholders in health disparities research initiatives. I have assumed thus far that genomics functions as a racial project of sorts, insofar as its understanding of race—as a diachronic, natural, immutable trait—undermines the enactment of social reform that health disparities research theoretically calls for. If health disparities research offers the opportunity to reform policy in such a way as to reduce the effects of racism, then the encroachment of genomics into this area of scientific inquiry is the most recent example of the legacy of eugenics. Recall that the argument for genomics research on breast cancer disparities was based in part on the assumption that the lived experience of race and its connection to health disparities had been fully and adequately investigated. This indirectly—but no less powerfully—discourages efforts to target poverty and discrimination as a way to work toward equality and the flourishing of all communities. As I discuss in the next chapter, genomics' conceptual indebtedness to the HGP, together with its subtle recuperation of the central dogma's logic of insularity and self-replication, necessitate a great deal of skepticism as to whether gene-environment research can be the progressive science its advocates envision it to be.

5

Genomics and the Polluted Body

In 2002 I attended a conference at Columbia University titled "Human Genetics and Environmental Justice: A Community Dialogue." Sponsored by the environmental justice organization West Harlem Environmental Action, the conference brought together geneticists, public health specialists, lawyers, and community activists to explain the science of genomics and discuss its social impact. It marked an important moment in the history of the environmental justice movement by having long-time activists directly engage a scientific discipline other than epidemiology. The gene-environment model informed the presentations of the speakers in a well-meaning attempt to elucidate the mechanisms of diseases like asthma that do not affect all groups equally. Nevertheless, I had to wonder: do we really need to know how gene variants predispose someone to have an adverse response to cockroach droppings? Shouldn't we just get rid of the cockroaches?

Indeed, why did West Harlem Environmental Action feel the need to hold such a conference? Since the sequencing of the human genome in the late 1990s, genomics has played an increasingly large role in US public health research. The Centers for Disease Control and Prevention (CDC) and the Environmental Protection Agency (EPA) both rely on the expertise of geneticists in order to classify risk among discrete populations of susceptible individuals. The CDC, in particular, considers the distribution of susceptibility alleles (genetic mutations that increase one's vulnerability to environmental toxins) to be a key area of public health research and policy.

This science, which I will call environmental genomics,[1] rests on several fundamental assumptions about heredity, risk, and the body. Taking breast cancer research as a case study, I argue in this chapter that environmental genomics research assumes that pollution is an inevitable feature of modern life, a naturalized feature that has no history, no politics, and no thinkable solutions—indeed, it ceases to be thought of as pollution at all. Instead, inherited genetic mutations are the polluting agents, a conceptual displacement that is part and parcel of a neoliberal logic that valorizes personalized, market-driven medical and public health interventions. Environmental genomics also understands race and gender to be diachronic attributes of the body, attributes that are in turn molecularized as mutations. This version of the polluted body is at odds with the discourse of environmental health and justice activism, which I describe here as offering a phenomenological perspective: pollution transgresses bodily boundaries, and this experience of pollution is what makes visible and politically meaningful these bodies' gender, race, and class markings. The embodied experience of pollution provides the ground for change, and in the name of social justice, not science.

Environmental Genomics and the Privatization of Risk

Before I present my argument about genomics, I must first briefly describe the public health context in which it is being introduced and differentiate mainstream public health from progressive alternatives in the work of social epidemiologists. Generally speaking, public health simply means the health of people—of particular populations, not individuals.[2] It entails, among other things, containment of contagious diseases, health promotion, and regulation of environmental pollutants. For the purposes of my argument, I focus on the latter, especially insofar as genomics is integrated into environmental health methodologies, although its influence is by no means confined to this aspect of public health.

The role of synthetic chemicals in industrial manufacturing and agriculture increased significantly after World War II, and with this increase came the need to test the chemicals' toxicity and regulate them accordingly. Despite passage of the Toxic Substances Control Act in 1976, however, most chemicals are not tested for safety before they go on the market in the United States. An oft-cited figure is that there are 80,000 or so synthetic chemicals on the market today, only 200 of which have been tested for safety.[3] In short, the United States regulates chemicals in such a way as to ensure that industry interests are protected to the greatest extent possible. The market largely controls what industry decides to use or not use in the manufacture of consumer goods and in agriculture.[4] This is in stark contrast to the situation in the European Union, which is much more likely to require safety testing before commercialization. Related to this is the widely accepted belief in the United States that it is up to consumers to force industry to replace toxic chemicals with safer alternatives. Lax environmental regulations, it is believed, can be circumvented through consumer-driven creation of markets for safer products.

An ethos of individualized, privatized fields of action ignores and thereby exacerbates the class, race, and gender dimensions of health disparities. Environmental justice activists have shown, for example, that dirty industries are concentrated in poor communities of color.[5] Citing evidence that dates back to the early 1970s, Robert Bullard[6] notes that a strong correlation exists between the conditions associated with political and economic disadvantage and disease.[7] An early study of the 1,000 residents (all of whom were black) of Triana, Alabama—a community contaminated with DDT and PCBs—revealed the highest human levels of DDT ever recorded.[8] Stress, lack of access to healthy foods, social isolation, and other factors make residents in such communities more vulnerable to diseases associated with exposure to environmental contaminants. Moreover, safer alternatives to many products are not available for purchase in these areas or are simply cost-prohibitive. In short, these communities are not well served by the pervasive and celebrated expansion of the market into public health and medicine, an expansion that is incompatible with the notion that social change, much of which can be effected at the policy level, is needed to correct the current state of affairs.

Although much attention has been paid in recent years to the problem of health disparities—for example, differences in the incidence of and mortality from breast cancer between white and black women—research in this field does not necessarily draw attention to the limits of the current regulatory structure and its increasing reliance on market principles. In part, this is because of a latent biologism that health disparities scholars can and do rely on. Biologism has, historically, worked in the service of free-market norms, explaining the social order of things as the natural order of things, belying the need to correct for the maldistribution of resources. Some health disparities research contributes to the biologism of race in particular and, by extension, its connection to class. As I argued in the last chapter, drawing on the work of Janet Shim and Merlin Chowkwanyun, when race is seen as a risk factor in epidemiological research, it remains insufficiently explained at best, susceptible to reification at worst. A pervasive lack of historical perspective in health disparities research strips race of its contingency and historicity; race reverts, it seems, to its biologistic, anachronistic origins. Reflecting a pragmatist, anti-theory orientation,[9] epidemiology on the whole pays too little attention to social causes or to the possibility that, as a discipline, it itself might be contributing to their obfuscation.

It is within this context that genomics is becoming increasingly important in structuring the thinking of environmental scientists, agency directors, and other thought leaders in the field of health policy. Broadly speaking, the turn to genomics has been motivated largely by the observation that even under similar environmental conditions (for example, similar exposures to a particular chemical), some people become sick while others do not. These genetically susceptible individuals (sometimes described as "hypersusceptible") are more prone to disease. Whereas disease-resistant people can metabolize chemicals in the environment without adverse effect (for example, because they harbor the genomic hardware necessary to repair the damage caused by carcinogens), genetically susceptible individuals cannot. Although high-penetrance mutations, such as those associated with BRCA1 and BRCA2 (sometimes called candidate genes),[10] have helped researchers explain a very small percentage of disease cases, they have little to no public health relevance. In contrast, polymorphisms—variations of genes that are present in more than 1 percent of the population—do. Advances in sequencing and information storage technology have enabled researchers to locate and catalogue just such genetic variation in human DNA.

Polymorphisms of genes do not in and of themselves cause disease, but researchers believe that many of them do play some sort of role in disease and disease resistance, either by themselves or in concert with other variants. They are "low penetrance" alleles, meaning that the risk they confer is slight. (In contrast, mutations of BRCA genes are considered "high penetrance," meaning they confer especially high risk—80 percent or higher for some women.) The cataloging of polymorphisms in human DNA allows researchers to study many variations simultaneously in order to determine how, collectively, they contribute to particular disease outcomes—one variation may not be significant in terms of penetrance, but may, in combination with others, markedly increase risk. Risk, therefore, is the cumulative effect of several polymorphisms of several genes synergistically interacting with environmental exposures. An example of polymorphism research is the HapMap project described in the previous chapter, a multinational effort to map groups of SNPs,[11] or haplotypes, that tend to be inherited unchanged. Because people share haplotypes, re-searchers can compare and contrast the haplotypes shared by those with a particular disease and by those without it. The identification of molecular differences in the haplotypes can then set the stage for determining just what role genetic variation plays in disease onset and resistance.[12]

Environmental genomics promises to elucidate the molecular mecha-nisms of variable risk in populations—the so-called gene-environment model. Environmental health advocates embrace this model, because they believe explanations of how environmental agents interact with the ge-nome in disease etiology can ground arguments for the ban or restricted use of individual chemicals or classes of chemicals. For example, Devra Davis, an epidemiologist and a fierce proponent of regulatory reform, has often invoked gene-environment interaction to demonstrate the causal links between exposure and disease. Some environmental breast cancer organizations, including those devoted to research and advocacy, take gene-environment research to be an essential part of an overall strategy to raise awareness about and politicize the connections between this disease and pollution. A newsletter of the Silent Spring Institute, for instance, in-cluded a discussion of research on PCB exposure among genetically sus-ceptible persons and the risk of breast cancer.[13] In short, environmental health researchers embrace the gene-environment model, believing that genetic variants are an important causal factor in complex biological processes. And policymakers believe that knowledge of genetic variation helps make prevention more accurate and cost-effective against a back-drop of scarce and dwindling resources for health care and public health. In 1999 Steven Coughlin, a researcher at the CDC, explained this reason-ing: "In the near future, the identification of genotype through genetic screening might allow for the identification of persons *truly* at high risk for an illness, targeted medical interventions, and improved allocation of health care resources" (emphasis added).[14]

The newfound popularity of the gene-environment model is reflected in several well-funded federal initiatives. At the CDC, the National Of-fice of Public Health Genomics[15] integrates developments in genomics with the CDC's public health mandate. Calling attention to the impor-tance of gene-environment interaction in health and disease, in 2001 the office's director, Muin J. Khoury, stated that "the public health impact of the human genome revolution is truly staggering." Common diseases and those responsible for "nine of the top 10 leading causes of death," he wrote, "have genetic components resulting from the interaction of genetic variation with modifiable risk factors."[16] Khoury has wholeheartedly embraced genomics research as a way for the CDC to devise more specialized, targeted screening programs. This latter mandate constitutes what the CDC sees as its role in "translational" research—taking knowledge of the genome and translating it into public health prevention. As Khoury put it, "in this genomics era, we need the entire research continuum, from gene discovery to development of practical tools for integrating genomics into population-based disease prevention programs. In this context, the applied public health research at CDC will evaluate what genes mean for health and disease in real communities in real time and, as importantly, how genomic information can be used to improve the public's health."[17]

At the National Institutes for Health (NIH), the Genes, Environment, and Health Initiative—launched in 2007 with an annual budget of $68 million, including $28 million in NIH funds already dedicated to similar projects—promoted environmental genomics in two ways.[18] First, it established a SNP database correlating gene variants with particular diseases, called the "dbGaP," or "database of Genotypes and Phenotypes." Second, it funded the Exposure Biology Program, led by the National Institute of Environmental Health Sciences (NIEHS), to develop new detection technologies to monitor individual exposure to environmental agents. These technologies generate data about exposures that can then be compared with data gleaned from analysis of biomarkers, such as blood and tissue levels of contaminants and somatically induced DNA damage.[19] The result is a determination of the thresholds beyond which exposures initiate biological changes that may lead to a specific disease outcome. Genetic variants are but one part of this exposure-disease continuum—they serve as biomarkers that allow diagnoticians to predict the likelihood that an exposure will lead to illness. NIH has touted the program as an innovative and long-awaited interdisciplinary approach to disease research, bringing together a significant number of geneticists and environmental scientists for the first time.[20]

Long before the Genes, Environment, and Health Initiative, the NIEHS had inaugurated research projects integrating genomics and environmental science. The Environmental Genome Project (EGP), now operated primarily at the University of Washington with NIEHS funds, was initiated to "better understand how individuals differ in their susceptibility to environmental agents and how these susceptibilities change over time." Like genomics research at the CDC and other programs at NIH, the EGP focused on cataloging SNPs for its GeneSNPs Database in order to encourage, and simplify, research linking specific diseases to gene variants. Taking a more targeted approach, the Environmental Genomics Group, also at the NIEHS, "works to characterize the role of genetic variation in human toxicological responses, especially to discover human alleles that modify responses to exposure and to investigate how such alleles affect risk in exposed people. This information is useful in determining appropriate variability parameters in human risk estimation models, in identifying at-risk individuals and in devising disease-prevention strategies."[21]

Whether these articulations of gene-environment interaction will lead to broad-based environmental regulatory reform depends on the ontological and epistemological status of the terms themselves. From a dialectical perspective,[22] DNA interacts with cellular components (such as proteins) in a mutually constituting process that, although limited by developmental possibilities, nevertheless allows for much contingency and variability. In this model, common terms associated with DNA, like "replicate," "code," and even "messenger"—as in messenger RNA, which by some accounts serves merely as a conduit for translation—no longer make sense. No one component of the biological organism may be said to have an identity or essence prior to its relational activity with other components.

In contrast, in the unidirectional model, best represented by Francis Crick's central dogma hypothesis, DNA is not dependent on environmental variables for its expression. Genes and environments interact with each other without ever fundamentally altering each other's substance. The genome's identity is understood to be stable over time, exerting its influence unilaterally. This understanding of the genome reduces the agency (and, by extension, the importance) of the material and natural environment. And it is a model that renders change at the production and social levels increasingly unthinkable. Like their eugenicist predecessors, postwar molecularists such as Crick believed that knowledge of the genome gave scientists unprecedented power to change it—genetic engineering became the sought-after scene of human ingenuity.[23]

Although this hubris is largely absent from genomics discourse today,[24] the antihistoricism and antimaterialism of the central dogma (and more recently of the HGP; see chapter 2 for an elaboration of this point) has nevertheless been reiterated in the rhetorical work of the gene-environment model, one that bears the conceptual and ideological traces of the central dogma and its privileging of nature over nurture as the field of action. Environmental genomics assumes, first, that pollution is an unavoidable feature of modern life and that only the genetically susceptible can and should be protected; second, that mutations and other biomarkers are the only source of meaningful, actionable knowledge; and finally, that the ideal subject of environmental health should be the individual who ultimately takes personal responsibility for his or her health—what amounts to the exercise of middle- and upper-class privilege.

Advocates of genetic testing are explicit in their belief that genomics can overcome the limits of current approaches to environmental regulation. A 2000 press release from the NIEHS claimed that genomics would "help businesses, workers, and local, state, and federal regulators fine-tune environmental policies to protect the most susceptible subpopulations at the lowest cost. Today . . . policy makers may over-regulate in hopes of protecting these extra-susceptible people—but cannot be sure the susceptible are indeed protected." More recently, in an essay laying out the rationale for environmental genomics, David Schwartz and Francis Collins claim that genomics makes risk assessment more accurate and cost-effective.[25] These statements reflect a common-sense, realpolitik approach to the environment that assumes regulation, in its current form, is inefficient at best and counterproductive at worst. Without knowledge of individual genetic variation, government agencies will presumably never know whether or not regulations will protect enough people, let alone the most vulnerable. Worse, the argument goes, regulations can and do inflict costly and potentially unnecessary burdens on industry.

The claim that regulation imposes unnecessary costs on industry, however, is both self-serving and needlessly conservative. At the very least, one must consider the degree to which it is based on evidence supplied by industry, evidence that at times it manufactured to serve its own interests. For example, apple growers in the early 1990s blamed public debate about the pesticide Alar for their devastating financial losses. The media largely accepted what amounted to a revisionist history, since the apple industry's financial problems preceded public evidence of Alar's toxicity, and producers were back on sound financial footing within five years of the controversy.[26] Indeed, some suggested at the time that the consumer backlash was the result of industry attempts to suppress the truth and/or lie about Alar.[27] Apple growers arguably had had enough lead time to phase out Alar and replace it with alternatives, as scientists, regulators, and growers alike were aware of the dangers Alar posed to children well before the public was. In short, industry has made and will always make the argument that regulations are too costly—it is reasonable to expect it to do so.[28]

Within this context, environmental genomics presents genetic subtyping of the population as the way to circumvent the need for broad-based environmental health initiatives. Yet the argument presumes that genetic analysis reveals more meaningful information than that produced by epidemiological methods. Genomic methods, however, do not reveal underlying categories of susceptibility so much as they define these categories in advance. And as is the case with genetic testing in general, there is a great deal of uncertainty around the interpretation of the results beyond simply placing someone in one of these predetermined subsets. The genetic test cannot specify the agent of action (for example, employer, employee, or policymaker) or say whether or not the person tested will actually succumb to disease if exposed. More important, genomics cannot say for certain whether or not those persons receiving negative results (that is, those who test negative for an allele connected to disease susceptibility) are really safe. Some of the most dangerous chemicals would, presumably, remain in use given the logic of the genomics model, provided those who show genetic evidence of susceptibility are removed from the exposure site. Take the example of benzene, one of the most dangerous chemicals in use today. Research has documented its harmful effects at exposures well below legal levels. We know this in part because of knowledge regarding the effects of benzene exposure at the molecular level.[29] The connection between genetic variants and the metabolism of benzene,[30] however, weakens the incentive to remove benzene from industrial sites,[31] thus permitting the exposure of most workers (that is, those who do not possess the genetic variants that we know affect metabolism). In a 2010 research article on benzene, for example, Qing Lan and colleagues admit to the "ubiquity" of benzene pollution and its economic importance to industry, thus providing a rationale for allowing the use of a chemical that was at one point deemed too dangerous at any level of exposure.[32] This logic is not confined to research on just one chemical—it informs the very federal programs that are setting the stage for a genomics-inflected approach to environmental and occupation health. "Many diseases are the outcome of a complex inter-relationship between multiple genetic and environmental factors," according to the NIEHS. "Research suggests that individual susceptibility is influenced more by certain genes than by exposure to environmental agents."[33]

And for those subject to testing, history does not offer many examples to recommend it. The intersection of undue industry influence and structural racism led to widespread discrimination in employment and education, and social stigma more generally, for African Americans in the 1970s and 1980s. The context was screening for sickle cell.[34] Being a carrier of the sickle cell gene—sickle cell trait—is much different from suffering from the disease sickle cell anemia. Nevertheless, despite little to no evidence linking carriers (that is, those with the trait, not the disease) with high altitude illnesses (aside from a few unexplained deaths of African American males in the US Army in the late 1960s), nearly all of the major airlines grounded or fired employees who tested positive for sickle cell in the early and mid-1970s.[35] One study concluded that African Americans with sickle cell trait were hypersusceptible to chemicals present in almost all industrial environments; as a result, DuPont excluded African Americans from many of its better paying industrial jobs.[36]

In the environmental health arena, the plausibility of the genomic narrative of risk is in many ways the outcome of long-standing critiques of epidemiological research methods. Although measurements of toxins in air and water cover a broad geographic area, they do not tell us about individual exposures—who has actually been exposed, how they were exposed (there are many routes of exposure—including through air, food, water, and consumer products), and whether the exposure has set off a chain of adverse biological events. Like other personalized exposure assessments such as biomonitoring,[37] genomics introduces a different spatial dimension to risk, moving from the environmental context of the body to the body itself. It also introduces a new temporal dimension, correcting for the failure of tissue screening (such as blood and urine analysis) to capture the dynamic history of exposure. Some chemicals metabolize quickly or are simply expelled from the body; cellular damage associated with the exposure (even prenatal exposures are thought to play a role in cumulative risk) will not necessarily be measured by examining bodily tissues for metabolites. These shortcomings have vexed epidemiologists searching for links between environmental pollution and breast cancer. Exposures to organochlorine chemicals like DDT cannot always be accurately quantified by looking at body burden. Breast-feeding, for example, rids the nursing mother's body of chemicals, rendering tissue analysis incomplete at best.[38]

The spatial and temporal dimensions of genomics-inflected environmental health casts biological markers as privileged sites wherein the histories of embodied experience of the natural and social environment intersect and condense. As Brenda Weis and coauthors wrote in 2005, risk is the outcome of "dynamic biologic processes in real time," what Schwartz

and Collins describe as the "individual and dynamic extent of the exposure and its impact on fundamental biological processes."[39] These markers
in turn isolate the field of action, a pragmatic need of public health and
medicine alike. According to Weis and coauthors, "the assessment of exposure and risk should focus on understanding the biologic processes of
human disease by defining markers that represent and link events, both
genetic and environmental, in the exposure-disease relationship."[40] These
novel methodologies will, in turn, permit early warnings or diagnoses and
chart a path for prevention and treatment.[41]

The privatization of risk is not unique to the encroachment of genomics
into environmental health.[42] Developing ways to ascertain evidence that
environmental exposures have occurred (for example, by detecting the
formation of DNA adducts, places where nucleotides have bonded with
chemical agents) arguably narrows the visual field to the body and invites
highly personalized interventions. Nevertheless, explaining this biological
activity with a hereditarian lexicon involves a different sort of biopolitics.
Environmentally induced changes are somatic; susceptibility is an inborn
trait. The heredity paradigm holds that genetic variants are insular entities, wielding their influence unidirectionally, unaltered by the biological
context. Schwartz and Collins, for example, write that the "assessment of
environmental contributions is much more difficult than for genetic ones.
The genome of an individual represents a bounded set of information, remains basically stable over time, and is very well suited to multiple analyt-
ical approaches."[43] In other words, susceptibility is a biological state that a
person inherits but cannot acquire; the genome is, for all intents and purposes, immune from the effects of embodied existence. Whereas epidemiology theoretically leaves open the possibility that disease is the outcome
of historical, political, and economic forces, the very idea of hereditary
susceptibility renders the historical-embodiment perspective, exemplified
by the approaches of social epistemology and political economy of health,
ineffective, even counterproductive. Schwartz and Collins, for instance,
point out that environmental exposures are difficult to isolate but genes
are not—genetic screening thus replaces the methodologically, politically,
and socially difficult process of gaining access to historical records, data
about toxic releases (which often take social movements and acts of legislation to procure), medical records, and the input of the public, all of
which are needed if the goal is to prevent or remediate environmentally
induced disease. Although, paradoxically, these environmental genomics advocates call for the transformation of environments, not genes,[44] an

inherent antihistoricism means that "environment" will be defined so narrowly as to be something of a person's own making.

The paradigm proposed by Schwartz and Collins is not altogether different—in theory or in practice—from that espoused by the central dogma enthusiasts of the mid-twentieth century, who claimed that hereditarian science had finally triumphed over the environmental models favored by progressive-minded scientists. At the time, this model of DNA served the institutional interests of geneticists in the United States as well as democratic capitalism more generally (see chapter 2 for a fuller discussion). Current environmental genomics rests on a gene-environment model that functions, rhetorically, to mask similar ideological work. It does so by calling into existence what Nikolas Rose and Paul Rabinow have called biosocial communities, so called because shared genetic markers are the source of a common identity that, in turn, underlies a range of bodily practices and identities. These practices are also, I want to suggest, characteristic of a neoliberal, entrepreneurial subject. By calling forth new communities of at-risk subjects who take personal responsibility for their health,[45] genomics discourse incentivizes new markets that serve the interests of the neoliberal order in two important, fundamental ways: first, by expanding the field of health consumerism (genetic susceptibility is a condition with which everyone, theoretically, must now contend and becomes the grounds for medical care and prevention) and, second, by legitimating the logic of market-based environmental regulations (personal responsibility and the mediation of genetic risk circumvents the need to eliminate the source of illness and risk). The biosociality called forth in environmental genomics discourse is thus an instance of how public health, as a mechanism of biopower, has served overlapping state interests in creating citizen-subjects whose duty is maximizing their health.[46]

It is important to note, however, that individualized care of the self involves a set of practices not necessarily available to everyone. And state regulation of health can also be a way to exert power over others. In the nineteenth century, for example, viral transmission was blamed on the susceptibility and thus inherent inferiority of ethnic and racial groups;[47] in more recent history, HIV/AIDS discourse has sanctioned fairly violent, homophobic policies and practices (in the "war" on AIDS, Catherine Waldby wrote, "the primary casualties . . . are . . . not the viruses themselves but people who are infected with the virus").[48]

The association of race, genetic susceptibility, and disease opens up the possibility of pathologizing the Other in altogether new and dangerous

ways—revealing that state actions are not necessarily designed to maximize the health of all.[49] For instance, although the stated *telos* of genomics research is individualized medicine, short-term reliance on racial
categories has produced new routes to racialization. The most common
methodology employed by environmental genomics programs relies
on maps of population-based genetic variation. The HapMap is often
cited—for example, by Schwartz and Collins—as a model for mapping
SNPs of interest; in fact, it has been recommended that data from the
HapMap be used, in the short term, to connect public health burdens
with specific genetic alleles.[50] Recall from chapter 4 that the HapMap
project, despite its stated purpose in mapping genetic variation across
the human genome, nevertheless selected as its representative sample
people identified as members of geographically isolated groups, groups
looking suspiciously like "races," anachronistically defined. The EGP
employed a similar methodology. Its GeneSNPs Database is based on
ninety-five individuals "representing the ethnic diversity found in the
United States." Although an improvement over the four "races" implicitly
employed by the HapMap designers, the database nevertheless opens the
door to associations between disease and ethnicity, associations that will
most likely not include whites, the dominant group. An additional way
in which racial and ethnic groups will be profiled will be by first defining
and targeting particular disease groups using the tools of environmental epidemiology. According to Weis and coauthors, "new study populations can be identified using global screening tools such as GIS-based
[geographic information system–based] technologies to identify specific
sub-populations with unusually high rates of the disease or potentially
elevated exposures for the disease."[51]

The obligatory consideration of race,[52] however, makes it likely that
the association between geography and race will ultimately be geneticized.[53] GIS defines a disease community by virtue of the inhabitants' location—understood as the multilayered confluence of a variety of social
demographic variables, such as education, income, and socioeconomic
status—in order for researchers to determine the role and influence of
confounding variables. Yet what happens when population alleles are included in the analysis? In a 2005 essay on GIS, lead, and children's health,
Marie Miranda and Dana Dolinoy describe how GIS can be used not only
to locate high-risk populations but also to incorporate gene-environmental models of lead levels in children's blood. Race, they say, "consistently
remains as an independent explanatory variable for elevated blood lead

levels" explained in part by "genetic polymorphisms that have race-based allele frequencies that have been linked to lead activity in the body."[54] In this example, the use of GIS does not so much capture the nuances of what it means to embody social identity in a particular place and time as it sets the stage for defining all mapped persons, regardless of their spatial and temporal location, as a biological race, defined by shared, inborn genetic susceptibility. Put another way, the spatial and temporal logics of genomics intersect with those of environmental science to produce a map of disease that relies on a diachronic, not a synchronic, understanding of embodiment, race, and disease.[55] A person's historically and socially determined experience of pollution is replaced by a hereditarian narrative—race becomes that which defines, a priori, the essential features of the body rather than that which can make sense out of, and in turn politicize, pollution practices.

The polluted body of environmental genomics is polluted insofar as it harbors genetic variants conferring susceptibility to environmentally induced disease. And it is these variants that are actionable: not in the sense of altering the constitution of one's DNA, but of modifying environments *on behalf of* this genomic field of constraint. Although the notion of human agency is present in this hereditarian narrative (genetic variants survive for a number of reasons, not just because they are the fittest), human history is, for the most part, condensed and ossified in the chemical base pairs of DNA. We may be compelled to intervene in our social worlds, but our agency is ultimately limited by the products of evolutionary history. Consider for a moment the oddity of using lead as an example of the necessity of both GIS technology and gene-environmental models of causation. Presumably lead, the toxicity of which is well established and for the most part uncontested, should be removed from children's environments regardless of genetic susceptibility. Miranda and Dolinoy essentially reopen the debate by raising the question of whether or not lead regulations benefit everyone.[56]

Breast Cancer and the Theory of Genetic Repair

In *Living Downstream: A Scientist's Personal Investigation of Cancer and the Environment*, Sandra Steingraber argues that cancer is a crisis of public health, an argument that departs from the lifestyle paradigm that has dominated the thinking of cancer researchers, the National Cancer Institute, and advocacy organizations like the American Cancer Society.[57] The lifestyle paradigm has been under fire from progressive researchers like Steingraber for its implied assumption that individual free will can trump the range of social factors that overdetermine the degree to which one is exposed to carcinogens—what Lewontin has described as "the social formations that determine the nature of our productive and consumptive lives." The lifestyle paradigm has raised questions that sound comical and intellectually bankrupt if asked in a different context.[58]

Steingraber's departure from the dominant paradigm of cancer research is theoretically and practically important: if cancer is understood as a symptom of historically specific industrial practices and a lack of regulatory infrastructure, the terrain of preventive measures shifts dramatically. More important, for Steingraber and other activists, the types of prevention that become thinkable within the public health framework they advocate for also stand the greatest chance of significantly lowering the incidence of cancer—a reduction that has remained elusive despite the spending of over $100 billion since the war on cancer was declared, well over forty years ago.[59]

Cancer rate data and epidemiological research together support the thesis that a significant percentage of cancers with rising incidence rates are caused by carcinogenic toxins in the environment. A look at the types of cancers with the highest rates of increase (and, in some cases, mortality) reveals a strong correlation with the widespread use of industrial and agricultural chemicals that began after World War II.[60] Cancers of the fatty tissues exhibit higher incidence rates than others, a phenomenon that may be explained by the fact that many of the chemicals introduced into the environment over the last fifty years tend to accumulate in the brain, breast, lymph nodes, bone marrow, and liver.[61]

Of the more than 200,000 breast cancer cases diagnosed annually in the United States, only an estimated 50 percent can be attributed to well-established risk factors such as reproductive history. This perplexing statistic has prompted reformers to call breast cancer an environmental disease, thereby making it an appropriate object of public health discourse. Indeed, there are several registers of evidence suggesting various environments are to blame: breast cancer rates vary according to geography and occupation, and they have been rising steadily—lifetime risk has nearly tripled over the last five decades, a rise that cannot be explained by the increase in mammography rates. Furthermore, women who immigrate to the United States have a greater risk for breast cancer than if they had remained in their countries of origin.

According to the Breast Cancer Fund's most recent "State of the Evidence" report, environmental agents with strong links to breast cancer include radiation, synthetic hormones, endocrine-disrupting compounds, and industrial carcinogens.[62] These environmental agents are found in X-rays, medical imaging technologies such as computed tomography scans (including mammograms), hormone replacement therapy, banned pesticides such as dieldrin, plastics, drugs, polyvinyl chloride, household cleaning products, solvents used in industrial production, and diesel exhaust.[63] A particularly intriguing theory is the "environmental endocrine hypothesis," [64] which holds that a diverse group of industrial and agricultural chemicals in contact with humans and wildlife can mimic or obstruct hormone function. In the case of breast cancer, the hypothesis potentially explains why exposure to chemicals that are estrogenic would play a part in rising rates, since estrogen is a well-studied risk factor for the disease.[65] One class of estrogenic chemicals, organochlorines, discussed in the previous chapter, are ubiquitous in the bodies of humans and wild animals; do not break down easily; and tend to "bioaccumulate"—that is, the amount of synthetic chemical body burden is much higher at the top of the food chain than at the bottom.[66] This explains why concentrations are high in large, predacious fish and particularly high in humans who depend on those fish for food. Synthetic chemicals that are classified as "persistent" (and thus likely to bioaccumulate) have been termed persistent organic pollutants. The United Nations has listed twelve of these pollutants as of greatest concern to scientists and environmental health activists.[67]

In April 1993 Mary Wolff and her colleagues published their finding that, after adjusting for known breast cancer risk factors such as age and reproductive history, women were four times more likely to contract breast cancer if they were exposed to significant amounts of DDT.[68] Although all study participants had some traces of DDE (the principal metabolite of DDT) and PCBs in their blood, the risk was greatest for those women showing the highest concentrations of DDT. Wolff and coauthors further caught the attention of researchers (and Congress) by suggesting that the findings had potentially profound implications for public policy. In the discussion section of the article they wrote: "Our observations provide important new evidence relating low-level environmental contamination with organochlorine residues to the risk of breast cancer in women. Given the widespread dissemination of organochlorines in the environment, these findings have immediate and far-reaching implications for public health intervention worldwide."[69]

Appropriating the language of Rachel Carson in an accompanying article titled "Pesticide Residues and Breast Cancer: The Harvest of a Silent Spring?," David Hunter and Karl Kelsey wrote that, given the age-adjusted rise in breast cancer, "it is somewhat surprising that few investigators have examined the relationship between pesticide exposure and breast cancer."[70] The findings of Wolff's team were given added weight in 1994, when Eric Dewailly and coauthors published their conclusion that women with estrogen-positive tumors had a higher body burden of DDE than women with estrogen-negative tumors, suggesting that certain types of breast cancer may be triggered by DDT exposure.[71] Three years later, Dewailly's team's investigation of the link between OCC exposure and estrogen-sensitive breast tumors prompted them to conclude that "the increase in estrogen receptor level in breast tumors observed during the past two decades could be explained in part by organochlorine exposure. If this hypothesis is confirmed, further limitations on production and usage of these chemicals will need to be implemented world-wide."[72] In a 1998 study of the pesticide dieldrin, researchers found that women with the highest traces of the pesticide were twice as likely to develop breast cancer as women with the lowest.[73] And more recent research suggests that age at exposure could be relevant: one study found that exposure at an early age to significant amounts of DDT is associated with increased risk of breast cancer later in life.[74]

At the same time as these studies established sound hypotheses regarding the connection between pollution and breast cancer, the identification of the BRCA genes was garnering an equal, if not greater amount of public notice.[75] After these genes had been sequenced and their contribution to breast cancer elucidated, it was not long before attention turned to "low" penetrance, more common genetic variants that contribute to breast cancer incidence.[76] With knowledge of these alleles, genomics could play a greater role in breast cancer prevention, in particular by parsing out the complex etiology of a cancer that afflicts nearly 200,000 women each year.[77] Making the case for the potential public health relevance of breast cancer genomics, one article noted: "The number of cases of breast cancer that are attributable to such genetic polymorphisms (in combination with environmental exposures) is likely to be much higher than the number of hereditary cases caused by mutations of high-penetrance genes such as BRCA1 and p53.[78] The genetic polymorphisms that may be linked to breast cancer are much more common in the population than are the high-penetrance cancer susceptibility genes."[79]

Described as "the longest-established environmental cause of human breast cancer,"[80] ionizing radiation (IR) has figured centrally in the development of a gene-environment model of breast cancer risk. The study of ataxia-telangiectasia (AT), an autosomal recessive trait,[81] revealed early links between genetic susceptibility to IR and breast cancer risk. People with AT exhibit hypersensitivity to IR (thus predisposing them to cancer) and suffer from a range of other diseases and disorders.[82] Heterozygous[83] carriers of the ATM gene ("ATM" stands for "AT mutated") are thought to be at greater risk for breast cancer than noncarriers.[84] Early research suggested that as many as 3.8 percent of breast cancer cases could be attributed to AT mutation heterozygosity; one estimate was 1–13 percent.[85] According to a 2006 review, ATM heterozygosity is a well-established breast cancer risk factor.[86]

The early association between the ATM gene and breast cancer risk coincided with (indeed, influenced) a research trajectory investigating the relationship between DNA repair and breast cancer risk from IR exposure, because the susceptibility that ATM carriers experience is due to the inability of repair genes to correct the damage from such exposure. In 2001 Eric Duell and coauthors examined effects of the repair gene XRCC1 on risk associated with IR and cigarette smoke. That study resulted in an important shift, as the variability of repair capacity in the general population (not just that subset afflicted with high-penetrance mutations of genes like ATM) became a topic of concern of researchers:

> Epidemiological studies using functional measurements of DNA repair suggest that DNA repair capability is variable within human populations. Because inactivating mutations in DNA repair genes are rare, it has been hypothesized that variation in DNA repair capability in the general population is a product of combinations of multiple alleles that show subtle variations in biological function. In support of this hypothesis, investigators at Lawrence Livermore National Laboratory recently discovered common variants in a large number of DNA repair genes, and it is proposed that those variants may act in combination with environmental factors to increase susceptibility to human cancer. A possible role for DNA repair deficiencies in cancer development has been the subject of increasing interest. In particular, previous studies suggested that breast cancer patients might be deficient in the repair of radiation-induced DNA damage. Several of these studies

reported reduced DNA repair capacity in family members of breast cancer cases, suggesting a potential genetic contribution to radiation sensitivity.[87]

Repair gene research thus provided a method of extending genomics research and explaining a much larger proportion of breast cancer cases.[88] Indeed, "The most rapidly growing research area is now focusing on the association between DNA repair genotype and phenotypes in human cancer susceptibility."[89] The relationship between DNA repair SNPs and breast cancer risk has been a growing area of study.[90]

There are many reasons why repair gene research should be regarded as a positive development in both the prevention and treatment of breast cancer. In terms of prevention, the types of assays developed by repair gene researchers can verify an individual's susceptibility to environmental mutagens well before damage has occurred. This is especially important for people for whom low levels of exposure are especially toxic. Repair gene research promises to shed light on the toxic effects of low exposures—not just to radiation, but also to other known or suspected carcinogens, such as environmental estrogens.[91] Extreme susceptibility to a particular carcinogen should warrant special protections at the workplace or home, as well as heightened surveillance. And as I observed earlier, understanding the mechanisms of carcinogenicity helps environmental health advocates' calls for stronger regulation of industry. Although the links between breast cancer and other toxins are not as well understood as that between the cancer and radiation, research regarding the latter has provided the evidence needed for environmental breast cancer groups to say that breast cancer is undoubtedly an environmental disease.[92] Repair gene research also contributes to breast cancer research in that knowledge of sensitivity to IR is crucial for women who are diagnosed with breast cancer and are considering radiation therapy as part of their treatment.[93] And such research may allow cancer doctors to develop interventions that will halt the proliferation of cancer cells by sensitizing them to radiation-based treatment. Simply put, by understanding how cells do or do not repair environmentally induced damage, cancer specialists could develop ways to target tumor cells and render them vulnerable to therapeutic interventions.[94]

Nevertheless, repair gene research invokes a model of gene-environment interaction that I argued earlier in the chapter is implicitly a privatized, market-based approach to public health. A guiding assumption of repair gene research is that the healthy person is one whose genes can repair the damage caused by exposures to toxic substances in the environment. In an article on repair genes and IR sensitivity published in 2002, Jennifer Hu and colleagues write: "Mammalian cells are constantly exposed to genotoxic agents. Distinct mechanisms have evolved to repair different types of DNA damage and to maintain genomic integrity."[95] Implicitly, industrial pollution becomes an unavoidable feature of modern life: the normal body is one that can repair the damage; the diseased body is one that cannot.[96] In this gene-environment model of breast cancer, pollution provides an opportunity for exploring mechanisms of DNA repair but is not, in and of itself, a target of intervention.

Indeed, the field of action shifts considerably in the discourse of repair gene research. This is possible because of the creation of new risk categories. Researchers hierarchically order those persons with the most efficient and effective repair systems and those with the least—what they call the "assessment of susceptibility genotypes."[97] This ordering does not so much assess these genotypes as it calls them into existence. In turn, these genotypes become new objects of public health surveillance. The bodies that occupy these new subject positions are, for all intents and purposes, polluted before being exposed to environmental toxins. A public health approach to breast cancer, according to the logic of this research, would entail identifying individuals on the basis of hereditary susceptibility. This is a significant remaking of public health practice: rather than identifying exposures first, then vulnerable publics (vulnerable because of their proximity to the site of the exposure, calling forth the importance of geography), environmental genomics would identify vulnerable populations first, then look at exposures. This temporal shift also means that only those exposures relevant to the hypersusceptibility genotypes will be considered for abatement.

The field of action also shifts in a second important way, and that is by calling into being new subjects of personalized, entrepreneurial health consciousness and practice. Genetic variants become the objects of public health discourse, a conceptual and material reductionism that abstracts the variant (and the body that harbors it) from historically specific social relations that overdetermine both the exposures and the body's responses to them. Hu and coauthors write: "To achieve effective cancer prevention, a better understanding of the carcinogenesis pathways initiated by DNA damage and the variations in DNA repair is required. Although exposure factors may be the major contributors to breast cancer risk, genetic susceptibility probably modifies their contribution at the individual level."[98] Exposures thus become meaningful (that is, actionable) only when individual responses to them are known; the ethical course of action is then modification of that person's environment. "One of the most important problems in preventing and controlling breast cancer," Hu and coauthors continue, "is the appropriate identification of high-risk individuals. Epidemiologic studies characterizing the molecular mechanisms of breast cancer risk are needed to further motivate genetically susceptible (sub)populations to get screened for early detection and intervention."[99] Environmental risk factors that are "amenable to intervention"[100] include diet and the ill-defined yet popular risk factor of lifestyle, which connotes both a privatization of risk and the neoliberal, consumer-culture values on which it depends.

The agent of disease—or, to put it another way, the agent of pollution— is the mutated gene or set of interrelated genes, not the industrial source of pollution. As a consequence, the agent of change is the individual harboring the genetic mutation, acting in concert with her physician. Mutations bear a historical narrative of evolution, making it difficult to see and comprehend the social situatedness of the woman with, or at risk for, breast cancer—put another way, her social location at the intersection of gender, race, and class.

This model of environmental health is intelligible only insofar as the gene occupies a privileged status of actionable knowledge, despite the inherent ambiguity of a genetic test. What if, for example, other genes or other biological factors are at play that may determine one's risk, factors that the test cannot reveal? Although understanding gene-environment interaction in the mechanism of disease is a laudable objective, it is a much more complicated process than genetic testing for susceptibility can speak to. The genetic test, as described by these researchers (and the directors of environmental genomics programs), presumably provides definitive evidence of inherited risk, evidence that can then inform and guide a set of individualized actions, which are, paradoxically, constrained by the immutable mark of heredity. To assume that a genetic test can and should set into motion one's modification of the personal environment is to embrace a reductionist thinking that is belied by the post-HGP revelation that human biology is much too complicated for any theory of DNA or genes as immutable modifiers of environmental exposures—all the more so when one takes into consideration the synergistic effects of many low-penetrance alleles.

In order to further understand the conceptual boundaries of genomics, consider an alternative approach. As I discussed in chapter 4, Martha Mc-Clintock and coauthors have shown how the experience of racism (including social isolation and hypervigilance) is associated with changes to the proper functioning of the BRCA genes (BRCA genes are thought to be repair genes, which explains their role in tumor suppression). For the McClintock team, the ideal approach to breast cancer prevention lies not in extending the heredity model to include low-penetrance alleles but in rethinking the relevance of heredity to public health—namely, by investigating environmentally induced genetic changes.[101] This, then, is genetic susceptibility taken to its logical (and perhaps progressive) conclusion: if low-penetrance alleles are common, and we are all at risk in some way, then the most plausible solution is to ignore SNP-disease associations and instead focus on otherwise healthy genes that are damaged by environmental exposures.

Genetic susceptibility categories conceptually and ideologically foreclose such an outcome, in part by opening up the possibility that genotypes will be racialized. Although the stated goal of breast cancer genomics research is individualized prevention for everyone, in the short term the individual-group paradox is inescapable. Repair research, dependent as it is on the concept of SNPs (enabling what Hu and coauthors call the "polygenic" model of breast carcinogenesis), inevitably racializes genetic variation and breast cancer risk.[102] This has occurred in a related research context: the genetic variation of metabolism of environmental substances. For example, a study on racial differences in estrogen metabolism, part of the Carolina Breast Cancer Study, hypothesized that racially stratified polymorphisms of the UGT1A1 gene could account for disparities in breast cancer incidence and mortality.[103] Drawing in part on the highly controversial claim that drug response trials have revealed racial differences,[104] the authors conclude:

> African-American women have been shown to be at increased risk for premenopausal breast cancer compared to white women. The risk of dying from breast cancer also differs by race, where a 20 percent increased risk is observed in black women compared to white women. Polymorphic variation within the UGT1A1 gene, especially the low activity alleles present at higher frequencies in blacks, is a possible genetic factor contributing to the observed differences. Supporting that observation, several studies report racial differences in drug effectiveness that have been partly explained by differences in drug metabolism . . .

Estrogen metabolism also appears to vary according to race, with a higher ratio of inactive:active metabolites in whites compared to blacks. Thus, it appears possible that polymorphic variations in UGTs[105] are one of the genetic factors accounting for these racial differences.[106]

Thus the racialization discourse analyzed in chapter 4 extends to the environmental health context as research shifts from improbable high-penetrance "founder" alleles of genes like BRCA1 and BRCA2 to low-penetrance genetic variations among women identifying themselves or being identified as black, African, or African American. Recall, though, that substantial evidence shows that social environments—themselves stratified by race and class—account for disease disparities (the likelihood of being diagnosed with premenopausal breast cancer is but one example). Chemical exposures as well as the experience of racism can alter biological processes, changes that can be mistaken as artifacts of heredity. Also as noted in chapter 4, socially induced stress can heighten the secretion of glucocortisoids, thus predisposing a woman to aggressive, premenopausal breast tumors. Moreover, environmental racism is another culprit in the ongoing investigation of health disparities in the United States. There is one noteworthy difference, however: whereas BRCA implicitly recast breast cancer disparities as hereditary phenomena for some black women (namely, young ones), polymorphism research, like that in the Carolina Breast Cancer Study, expands the heredity hypothesis to possibly include all black women.

Although there is some chance that the epidemiological research exemplified by McClintock and others will resist geneticization, I have also argued throughout this chapter that this research, too, may be constrained by the power of proximate causes to capture the imaginations of doctors and public health officials. We may need a biosociality not only without genes, but without biology. In the next section, I explore this idea by discussing three areas of environmental health social movement: environmental justice, biomonitoring activism, and environmental breast cancer.

Race, Class, Gender, and the Embodiment of Pollution
Environmental Justice

Over thirty years in the making, the environmental justice movement has united its members around the shared experience of pollution. Notably, this movement does not rely or depend on elucidation of the biological mechanisms (inherited or acquired) underlying pollution exposure and disease. Rather, activists have engaged in political and social activity on the grounds that it is unethical to disproportionately expose entire groups of people to toxic chemicals.[107]

Much evidence suggests that racism explains both the concentration of dirty industry and the relative ease with which corporations violate environmental regulations. Several studies released throughout the 1980s and early 1990s showed that the greater toxic burden on African Americans was not entirely attributable to income and status. In 1983 the Government Accountability Office released a study of hazardous waste sites in the southeastern United States, reporting that the majority of people in three out of four areas studied were African American.[108] (Emelle, Alabama, where the nation's largest hazardous waste landfill is located, is 78.9 percent African American.) In 1987 a study conducted by the Commission for Racial Justice confirmed the findings of the Government Accountability Office.[109] And in 1992 a study published in the *National Law Journal* concluded that penalties against those who broke pollution laws over a seven-year period were lower if the violation occurred in a minority area: the average penalty was $335,000 in white areas but only $55,000 in minority areas. The study also found that it took, on average, 20 percent longer to place hazardous waste sites in minority neighborhoods on the Superfund National Priority clean-up list. The EPA has also taken longer to act in minority areas even when they are declared Superfund sites and has been more likely to order containment instead of treatment. This was the case regardless of whether the community was wealthy or poor.[110]

The collective experience of what environmental justice activists have termed "environmental racism" calls for the historical study of both "race" and "racism." Race, the embodied experience of racial identity, intersects with location, political economy, and class, all of which must be studied synchronically. For example, the South has been home to an inordinate amount of industrial pollution, especially in impoverished minority communities. Much of the Gulf Coast region was industrialized in the 1970s and 1980s as a result of its low-wage, nonunionized labor force; access to natural resources; tax breaks; and weak environmental regulation and enforcement. Moreover, corporations took advantage of the region's need for economic development by promising that industrialization would solve persistent problems of unemployment and poverty. Louisiana in particular has had the unfortunate privilege of being the preferred site for the petrochemical industry.[111] Like other southern states, Louisiana is rich in natural resources (for example, its salt domes have been used for storing oil; brine, also plentiful, is necessary for the production of chlorine, which is used in the production of plastics)[112] and is hostile to organized labor. In fact, the chemical industry increased production in Louisiana after it faced significant labor opposition in Texas in the 1950s.[113] A study commissioned by Greenpeace in 1988 found that East Baton Rouge Parish had "more violators of emissions permits, commercial toxic waste facilities, employees in petrochemicals, and toxic waste generation than any other county along the [Mississippi] River, and, in addition, ranked second or third for 6 toxic emissions measures, 3 toxic discharges measures, and for toxic waste landfills and incinerators."[114]

Geography alone does not fully explain these patterns—evidence surfaced in the mid-1980s that industry strategically targeted communities marked by poverty and political disenfranchisement.[115] Industry also sought communities with documented antiregulation attitudes; it was effectively looking for communities with no history of economic justice social movements, itself a sign of political alienation. A 1984 study prepared by the consulting firm Cerrell Associates for the California Waste Management Board profiled neighborhoods most likely to successfully fight waste disposal sites. The study's authors observed that "all socioeconomic groupings tend to resent nearby siting of major facilities, but middle and upper socioeconomic strata possess better resources to effectuate their opposition. Middle and higher socioeconomic strata neighborhoods should not fall within the one mile and five mile radius of the proposed site."[116] The Cerrell report became "the handbook for site location in the toxic-waste industry"[117] and provided the following criteria for possible industrial development: residents should not be property owners; they should have a "free market orientation" and should not appear to favor a "socialist-welfare state"; and they should be employed by "nature exploitative occupations."[118] Industry counted on the political powerlessness of the mostly poor, African American population, virtually all of whom were deprived of the right to vote. By concentrating refineries and other factories in these communities, industry gained access to cheap land without worrying about political opposition.[119] Once a site is chosen, the study concluded, industry should "decide, announce, defend."[120]

The case of Convent, Louisiana, shows how devastating this pattern of industialization has been. In 2002 the population was 21,000 people, of whom 51 percent could not read or write, 61 percent were unemployed,

and over 65 percent lived below the poverty line.[121] Of the twelve industrial plants operating within the thirty square miles centered on Convent,
nine are chemical plants, and nine are listed as major sources of toxics
releases.[122] This dense pattern of industrial production has resulted in
astounding amounts of toxic pollution: "While on average, 7 pounds of
toxic materials were released nationwide into the air for every person
living in the United States as a whole, 2,227 pounds of pollutants were
released into the air for every person living near Convent."[123] According
to other estimates, the release rate is even higher: although the average
urban releases are 10 pounds a day per person nationwide, in Convent the
figure is 4,000.[124] Researchers have observed a very high cancer rate, especially breast, ovarian, testicular, and scrotal cancer. Many residents also
suffer from pancreatic cancer, which is rare elsewhere. And young girls
are at considerable risk of endometrial cancer.[125]

Recognizing that political economy, racism, and geography explain
these releases, residents politicized their experiential knowledge of bodily
suffering. The first national environmental justice protest took place in
1982 in Warren County, North Carolina, in opposition to the proposed
burial of 32,000 cubic yards of soil contaminated with PCBs in a mostly
black American community,[126] leading to the first-ever arrests (400 altogether) for opposing a toxic waste site.[127] A year later, the Urban Environment Conference in New Orleans brought together representatives of
labor, civil rights, and environmental groups for one of the first meetings
ever organized around the issue of environmental justice.[128] In 1991 environmental justice activists attended the inaugural national conference
on the issue, the First National People of Color Environmental Leadership Summit, in Washington, D.C. (the second summit was held in 2002).
Attendees in 1991 represented a variety of organizations, including those
concerned with civil rights, community development, public health, and
environmental protection.

Not surprisingly, Louisiana would be the site of one of the first efforts
to invoke federal law in the name of environmental justice.[129] It was also
the site of a historic environmental justice campaign against a major corporation. In the late 1990s, Shintech, a subsidiary of the firm Shin-Etsu,
proposed a huge, $700-million PVC plant in Convent,[130] just a few miles
from two other petrochemical plants. Residents of Convent refused to accept the argument that this mode of economic development would actually benefit the community.[131] The Louisiana Coalition for Tax Justice
estimated that the tax exemptions promised to the company—which

included an exemption from local and parish property taxes for five years with a possible five-year extension, plus tax credits for each new job created—would amount to more than $94 million over ten years, an amount that would include $27 million in lost school tax revenue.[132]

Activists took this and other information to the community and beyond, going door to door, leading toxic tours, organizing letter-writing and petition campaigns, and holding town meetings. They also engaged in participatory research, gathering crucial demographic and epidemiological data. As a result, for the first time in the nation's history, the EPA granted a suspension request due to suspected environmental racism.[133] Before the EPA could finish its investigation, "the combination of protests, legal actions by Tulane's law clinic and community groups, ongoing negative publicity, and the threat of a precedent-making federal action finally caused Shintech in September 1998 to withdraw its plan to build the plant at the Convent site."[134]

Although the goal of the environmental justice movement in Lousiana and elsewhere has been the amelioration and prevention of pollution and disease in specific areas, its linkage with civil rights and economic justice has allowed it to demystify one of the more devastating ways in which institutional racism exacts its effects. These residents are not so much chemically exposed as they are chemically assaulted.[135] This social movement has also provided the opportunity for coalition building[136] and a vision of social transformation that includes long-term, primary prevention of environmental disease and its differential effects. Indeed, the obstacles faced by opposition groups and the successful siting of industrial facilities in the face of this opposition[137] demonstrate the need for, at the very least, regional economic justice and political empowerment. For example, activists in Louisiana have struggled (not always successfully) to overcome state, municipal, and business support for chemical manufacturing as well as relentless attacks on their credibility by well-funded, influential elites.[138] In 1998 the Louisiana Supreme Court ruled that, among other things, law students could no longer represent clients or groups affiliated with national organizations (the Tulane Environmental Law Clinic had provided pro bono assistance to activists), and that the majority of the students' clients (75 percent) must satisfy the definition of "indigent."[139] Recalling a hearing during which one student lawyer defeated at least eight Shintech attorneys, the law clinic's former director, Robert Kuehn, said: "The sad commentary is that the governor and the court are trying to say that eight licensed attorneys against one student attorney are not good enough odds. They now want to make it 8 to 0."[140] The local press characterized the assault as nothing less than "class warfare."[141]

Social movements, grounded in an understanding of race as the synchronic and embodied experience of social relations, further act as a counterweight to one of the most powerful weapons of elites: scientific expertise. It is very difficult to provide definitive proof of a cause-and-effect relationship between exposure and disease—indeed, the strategic use of scientific evidence (or lack thereof) by industry is well documented.[142] Science has raised awareness of the profound impact that environmental racism has on the bodies and lives of its victims (as I noted earlier, some activism did involve collecting raw data about disease types and rates), but ultimately it cannot be the ground for action given the unavoidable uncertainty that is part and parcel of the scientific method. Simply put, there will always be some studies that undermine even the most progressive, participatory research. In most cases, efforts to improve public and environmental health move forward because they are the ethical and prudent thing to do, not because there is conclusive scientific evidence demonstrating a link between a particular chemical and an adverse health effect.[143]

Environmental justice activism in particular has shown that the successful movement from experiential knowledge[144] to direct action and policy change does not require epidemiological evidence of cause and effect between exposures and particular disease outcomes. The call for change has been a call for justice, not merely scientific understanding and medical care. And it is a call for justice based on knowledge of the historical and racialized patterns of industrial growth and the state and corporate practices responsible for them. By making the undue suffering of disease, death, and poor quality of life a civil and human rights issue, embodied experience is politicized. Environmental justice has drawn attention to the historical forces and geographical specificities of strategically deployed pollution—making racism, not race, the central issue.

Biomonitoring and the Politics of Body Burden

Another method for mapping the histories and geographies of exposure is through biomonitoring, or the testing of urine, blood, and breast milk for chemicals and their metabolites. It has demonstrated, quite alarmingly, the ubiquity of synthetic chemicals—many of them toxic—that we absorb from the air we breathe, the food we eat, the water we drink, and the products we handle. The chemicals in question have been implicated in a variety of chronic diseases and reproductive disorders.

In 2001 the CDC's National Center for Environmental Health released its first "National Report on Human Exposure to Environmental Chemicals," which made public, for the first time, data from the CDC's National Biomonitoring Program. Interest in implementing biomonitoring programs since this first report has been substantial. In 2005 California enacted the country's first statewide biomonitoring program—the California Environmental Contaminant Biomonitoring Project—and many other states are considering similar programs. Nonprofit organizations such as the Environmental Working Group have made biomonitoring studies central to their advocacy efforts, and community groups around the country have conducted biomonitoring studies as a way to raise awareness about the vulnerability of the human body to chemical exposures.

These projects provide evidence of the ubiquity of chemical pollution, its unpredictable movement, and the permeability of body boundaries. Importantly, they need not be either disease-driven or dependent on the organic experience of suffering.[145] They are epistemologically and ontologically significant because they provide material evidence of vulnerability; this newly perceived permeability of body boundaries reveals unforeseen connections or relations with other bodies in local, national, and even global contexts. Biomonitoring is a form of social action that can potentially effect significant policy and social change without recourse to knowledge of shared genes. And finally, it shows how progressive change can be realized within a particular milieu—in this case, "risk society."[146]

Among the many features of risk society, according to Ulrich Beck's landmark study, is the nature of risk itself. Some of the most pressing environmental risks we face, Beck says, are largely undetectable—they have no taste, smell, or color.[147] Second, risk is constructed: it does not exist prior to the discourse that constitutes it as such. It is a term lacking a referent, a term without a corresponding material object. Together, these features of risk have entailed a certain set of relations between the public, environment, and knowledge. Beck explains:

Risks induce systematic and often irreversible harm, generally remain invisible, are based on causal interpretations, and thus initially only exist in terms of the (scientific or anti-scientific) knowledge about them. They can thus be changed, magnified, dramatized or minimized within knowledge, and to that extent they are particularly open to social definition and construction. Hence the mass media and the scientific and legal professions in charge of defining risks become key social and political positions.[148]

Consider one particular dimension of the risk society as it has emerged in the US context: the rise of risk assessment. Risk assessment predicts the number of deaths and/or diseases expected to result from a given exposure. Cost-benefit analysis, which is closely linked to risk assessment, is the procedure by which government and industry weigh the number of expected deaths and diseases against the cost to business from regulation and/or the economic cost to a community where the industry in question is or will be located. Risk assessment gained considerable traction when, in the midst of several conflict of interest investigations of administration officials, then-President Ronald Reagan selected William Ruckelshaus to head the EPA. Ruckelshaus championed cost-benefit analysis and risk assessment in an effort to make decisions at the EPA appear more objective and scientific. Risk assessment also benefited from the Supreme Court's 1980 decision in *Industrial Union Department, AFL-CIO v. American Petroleum Institute* regarding the benzene exposure standard of the Occupational Safety and Health Administration (see endnote 32), a decision that threw out the requirement that government agencies keep exposure levels of known carcinogens as low as technically feasible (the standard under question had been based on the assumption that there are no safe doses of carcinogens). Shortly after the decision, Reagan issued an executive order requiring cost-benefit analysis for all major environmental regulations, providing more impetus for the hegemony of quantitative risk assessment. Unprecedented power was given to the Office of Management and Budget to reject or delay regulations if their cost to industry was deemed too high.

Risk assessment is now used to regulate a variety of media: pesticide residues in food, air pollution, carcinogens in drinking water, and Superfund waste sites. Much of risk assessment's legitimacy extends from the emergence of coextensive specialized discourses of risk expertise, reflecting what Beck termed the "theoretical" or "scientized" consciousness of risk society.[149] With the proliferation of risk assessment, risk experts have come to play important mediating roles: they are employed by government agencies and privately funded think tanks (for example, the American Enterprise Institute has championed risk assessment for some time); they hold prestigious posts in academia; and they enter into the public sphere when environmental science writers call on them. Risk experts in many ways occupy important positions in the "professional-managerial class,"[150] whose interests overlap those of the upper classes even if their scientized discourse mystifies this convergence.[151] Indeed, the existence of shared interests is by no means transparent—experts seemingly do not side with industry or with citizens. They instead occupy an apparently objective, value-free middle ground from which they mediate controversy. But risk experts accomplish this by operating within what has been called the "deficit-model" of public knowledge and understanding of science[152]—a strategic, self-serving, and not altogether accurate construction of their audience.[153]

Risk assessment and cost-benefit analysis entail two fundamental assumptions that have been the source of contention among community activists and academics alike. First, they assume that the notion of "acceptable risk" is a value-free, objective determination. Critics of risk have countered that "acceptable" is inherently normative, generally reflecting industry interests only, masquerading as common sense. Second, and related to this, risk assessment and cost-benefit analysis assume that risk should be managed, not reduced or eliminated. However, "managing" risks misdirects public debate about risk away from the search for safer alternatives and prevention more generally.

Biomonitoring projects are not necessarily immune from these criticisms. They do, after all, produce data that can be translated into familiar frameworks of risk assessment: a person may test positive for traces of a particular chemical, but that does not necessarily mean that researchers or physicians can say for certain whether those detectable traces pose any sort of health risk. Nevertheless, biomonitoring projects have the potential to disrupt dominant risk narratives, and in many ways they have already done so. Similar to environmental justice's implicit critique of risk assessment—in particular, the latter's reliance on scientifically ascertained causes and effects and the notion of "acceptable risk"—body burden data motivate both social movement organizing and policy reform regardless of whether research demonstrates a connection between detectable exposures and disease. Biomonitoring achieves this by changing the way we think about our relationship with industrial practices and the free market principles that largely govern them. Biomonitoring studies compel participants to experience pollution as embodied and, in so doing, connect our lived experience with the lived experience of others—and with the social structures responsible for its production. In part, this is because the data are always already culturally inscribed, as are the tissues and fluids used to determine the degree of chemical exposure. Breast milk comes from female breasts, which are themselves gendered. This raises the possibility that biomonitoring can effect the experience of pollution as gendered, thereby politicizing it in specific and powerful ways. [154]

Biomonitoring asks us to rethink not just epidemiology, toxicology, and public policy, but also social theory itself. For example, the collection of body burden data reverses the invisibility of risk that, according to Beck, characterizes modern society. Although the data do not in and of themselves provide evidence of risk, the presence of these toxins nevertheless provides a type of visual evidence of the permeability (and violation) of body boundaries and of the largely involuntary nature of exposures.[155] Thus, even though some advocates of biomonitoring strategically employ the rhetoric of scientific innovation and present body burden as a new, unique type of evidence that can improve epidemiological assessments,[156] action is not necessarily predicated on scientifically ascertained cause and effect and calculations of acceptable risk. Action is rooted in the knowledge that toxins are present even if not felt, and that the source of exposure should be eliminated on ethical grounds.

In sum, biomonitoring makes possible a different kind of body politics, one in which the body is a text with which to read and interpret industrial practices and their transformation of the natural world. The narratives that are made possible are ones in which bodies are part of the environmental and social landscape; it is a synchronic history and a phenomenology of pollution, one in which gender, race, and class become meaningful after knowledge is transformed into action.

Environmental Breast Cancer

The environmental breast cancer movement, a national movement some twenty years in the making, has transformed the disease from a lifestyle affliction to a matter of public health. A closer examination of this movement reveals further a body politics that has made breast cancer a matter of gender justice more broadly.[157]

Consider, for example, a striking feature of many environmental epidemiology studies on organochlorine compounds and breast cancer risk: *all* samples show measurable levels of these chemicals. Epidemiologists must therefore calculate risk by comparing women with the highest levels of body burden and those with the lowest. There simply is no control group to be found, because all women have been involuntarily exposed to toxic pollutants and, once exposed, store chemical metabolites in the body's fatty tissues—in this case, the breasts.

The act of breast-feeding reveals the biologically complex and thoroughly gendered dimensions of body burden.[158] Testing over the years has revealed both the permeability of body boundaries and the persistence of chemicals classified as organochlorines. Many synthetic chemicals bioaccumulate (become concentrated) in human breast milk, making it one of the most contaminated foods on the planet.[159] Richard Jackson, former director of the National Center for Environmental Health at the CDC, expressed the concern that "if you compared the legally allowable levels of toxic chemicals in human milk, you would conclude that half the human milk in the United States would be unfit for sale under FDA [Food and Drug Administration] guidelines."[160] As I observed earlier in this chapter, one way in which a woman can rid her body of these pollutants is through breast-feeding, thus casting into sharp relief the ways in which women's bodies are inextricably tied to others.[161] And as Steingraber notes in *Having Faith: An Ecologist's Journey to Motherhood*, breast-fed infants, not their mothers, are at the very top of the food chain, relying on milk containing the largest amounts of bioacculumated toxins.[162]

The body burden of chemical contaminants—its history, geography, and modes of exchange—casts gender in quite a different light than has been typical in scientific research on breast cancer.[163] Consider the risk factors for breast cancer most likely to be recited in public and specialized discourses alike: reproductive history, ethnicity, and socioeconomic status, the latter two being more or less substitutes for reproductive history. In practice, all three understand gender as an identity bound up with reproduction and thus reducible to it. Contrary to this, the body burden data in environmental breast cancer research implicitly presumes gender to be a property of embodied existence. It thus cannot be reduced to reproduction but instead reflects, and in turn imputes meaning to, the experience of living as a gendered person within a particular set of social relations—themselves inextricably tied to specific modes of production.

For example, environmental breast cancer research shows the importance of understanding disease rates regionally and historically. Although methodological limitations make it difficult to ascertain when an exposure took place, the route it took, and the amount of chemical involved (not to mention the synergistic effects of exposure to multiple chemical agents), this research nevertheless raises the novel question of the importance of a particular kind of historical and contextual knowledge in the pursuit of scientific understanding of disease. Regionally, exposures have varied, thus illustrating how breast cancer is tied to practices beyond the reproductive context. Research from the Silent Spring Institute, for example, has focused on breast cancer rates and causes in specific regions of Massachusetts, using a GIS to track variations on Cape Cod (historically known for higher-than-usual breast cancer rates, in part, it is thought, due to pesticides used in the area's cranberry bogs), urine samples to determine body burden, bioassays to measure the estrogenicity of particular chemicals, and epidemiological research to investigate risk factors associated with socioeconomic factors such as domestic chemical consumption (thereby qualifying socioeconomic status in an interesting way, given that higher education and income are associated with more chemical exposures—for example, those stemming from household pesticides and dry cleaning). This research directly challenges the implicit tie between modernity, feminism, and cancer that subtends mainstream breast cancer risk discourse, including the claim that rates are explained by longer life spans, better detection technologies, and having fewer children later in life.

Environmental breast cancer activists have repeatedly challenged the claim that well-established risk factors explain regional variations in breast cancer statistics. In Long Island, for example, the New York State Department of Health, in collaboration with the Nassau County Board of Health and the State University at Stony Brook, published the 1990 "Long Island Breast Cancer Study."[164] It exonerated environmental pollution as the primary suspect for high breast cancer rates in Suffolk and Nassau Counties, the two counties where self-made cancer maps rallied residents around the issue of environmental breast cancer. Using a questionnaire method, the authors of the study concluded that rates were explained by the established risk factors of Jewish ethnicity,[165] socioeconomic status, diets high in fat, and having children later in life.

The state's study results did not deter activists, who responded by stepping up their organizing efforts considerably. In 1990 the organization One in Nine was formed,[166] cofounded by Fran Kritchek, the long-time activist Barbara Balaban,[167] and Marie Quinn. These and other activists recognized that there was room for "tentative optimism" that polluted drinking water and proximity to landfills could be linked to high breast cancer rates, and they were not dissuaded by health officials' conclusion that no further research was needed because established risk factors offered a satisfactory explanation.[168] Experiential knowledge by residents challenged the assumptions of breast cancer researchers, and cancer maps revealed that breast cancer was also high in poor and working-class neighborhoods.[169] Women attending support group meetings at Adelphi University, for example, were not disproportionately old, Jewish, or rich. As Balaban noted, "the women were getting younger compared to the previous years."[170] However, a second study by the CDC affirmed the state health department's conclusion that breast cancer incidence could be explained by Long Island's demographics. According to the CDC's researchers, when risk factors such as reproductive history, socioeconomic status, and ethnicity were taken into account, the risk was not higher for Long Island residents.[171]

Activists continued to mobilize, challenging the notion that women's ethnicity or income and profession should be studied without taking environmental variables into consideration. Their efforts led to the June 10, 1993, passage of the federal Public Law 103-43 and with it the creation of the Long Island Breast Cancer Study Project (LIBCSP), the mandate of which was to study "potential environmental and other risks contributing in [sic] the incidence of breast cancer in Nassau and Suffolk counties" and the two counties with the highest age-adjusted mortality rates for breast cancer during 1983–87, which were Schoharie County, in New York, and Tolland County, in Connecticut. In addition, the law directed that the study "shall include the use of a geographic system to evaluate the current and past exposure of individuals, including direct monitoring and cumulative estimates of exposure, from (1) contaminated drinking water; (2) sources of indoor and ambient air pollution, including emissions from aircraft; (3) electromagnetic fields; (4) pesticides, and other toxic chemicals; (5) hazardous and municipal waste; and (6) such other factors as the [National Cancer Institute] director determines to be appropriate."[172] Funded and coordinated by the National Cancer Institute (NCI) and the NIEHS, the LIBCSP included a variety of research projects: epidemiological studies, a breast and ovarian cancer registry, and laboratory research examining the biological mechanisms by which synthetic chemicals might cause breast cancer. At the time, the LIBCSP was described as "ground-breaking epidemiological research"[173] with the "most power to look at the relationship" between breast cancer and toxic chemicals.[174] In addition, the LIBCSP is a rare example of a research program resulting from an act of Congress (Congress generally provides funding for agencies like the NCI, not for specific projects). Lawmakers further deviated from traditional protocol by directing the NCI to pursue specific research questions and employ specific methodologies (such as GIS) and by requiring public participation in the research process (this included holding regular town hall–style meetings with residents).

The LIBCSP demonstrates one way in which social movement activity is important for breast cancer prevention: it helps promote scientific innovation. Following the example of successful HIV/AIDS activism,[175] these breast cancer advocates fought for and secured a "seat at the table," assuming responsibility for raising research funds, gathering data (hence the term "popular epidemiology"),[176] designing studies, and serving on advisory panels. Social movement theory suggests, in fact, that activism results in better epidemiological methods.[177] Cancer registries, right-to-know legislation, and biomonitoring programs are essential to researchers, because sound and adequate information enables more sophisticated studies leading to conclusions with higher levels of confidence. The following anecdotes support this belief: Years after witnessing breast cancer cells in culture reproduce as if they had been exposed to the hormone estrogen (also know as estradiol), the Tufts University researchers Ana Soto and Carlos Sonnenschein ascertained that nonylphenol, a chemical added to plastic to make it flexible, had leached from laboratory materials into the cultures. It took them a considerable amount of time to prove this because the manufacturer of the plastic refused to disclose the list of chemicals used in its production, claiming proprietary privilege. Similarly, David Feldman and Aruna V. Krishman spent more than a decade trying to figure out how and why a form of yeast produced estrogen. In this case, bisphenol A, a product of the breaking down of polycarbonate (used in many plastics), had leached into the yeast cultures after they were exposed to high temperatures.[178] Had right-to-know legislation existed in the states in which these researchers worked, they may have been able to explain these anomalies much sooner.

Nevertheless, the environmental breast cancer movement also illustrates, quite compellingly, the need for a sustained, broad-based social movement that includes grass-roots organizing and consciousness-raising, lobbying for legislative reform, and direct action based on the conviction (similar to that espoused by environmental justice and biomonitoring advocates) that unfettered releases of toxic chemicals are unethical. It is, simply put, advocacy that does not necessarily depend on conclusive evidence of cause and effect. Environmental breast cancer activists were early supporters of the precautionary principle, a policy norm placing the burden of proof on manufacturers so that they must demonstrate the safety of synthetic chemicals before they enter the marketplace. The precautionary principle further calls for regulation of these chemicals to be based not on incontrovertible scientific evidence (recognizing that scientific "certainty" is a mere construct and rarely, if ever, realized), but on evidence strongly suggesting the potential of synthetic chemicals to significantly compromise human health.[179] Moreover, as Maren Klawiter observes in her ethnographic study,[180] some environmental breast cancer activists have been justifiably suspicious of science-driven activism—extending their suspicions even to the practice of participatory research. These activists have rightly questioned whether epidemiology can produce actionable evidence of risk, given the inherent uncertainty of epidemiological research and the ways in which science funded by industry (usually via trade associations) permits it to strategically employ arguments about risk to delay environmental regulation. Demands for environmental reform, these activists believe, should never be contingent on the work of epidemiologists, regardless of their intent.

Controversy around the organochlorine hypothesis is just one important example of the need to tie arguments about environmental breast cancer prevention to justice and ethics, not science. Notably, several studies—including those of the LIBCSP—failed to confirm a link between exposure to organochlorines and breast cancer risk.[181] Science journalists and researchers alike have used the doubt generated by these studies to argue against further research. In 1994, for example, the *Journal of the National Cancer Institute* published a letter to the editor from Stephen Sternberg, arguing that the results of a single study proved that breast cancer is not a public health issue. Nancy Krieger, the author of said study, later accused Sternberg of propagating antiscientific and anti–public health sentiment. (Krieger argued that the spirit of "science" is continuing inquiry, regardless of conflicting results.)[182] This exchange, and others like it, relegated to the technical sphere, had little positive effect on public discourse about the organochlorine theory at the time.

Evidence supporting the environmental breast cancer hypothesis continues to accumulate, granting it ever more credence. Nevertheless, I want to suggest here that the environmental breast cancer movement remains a strong, viable, and diverse voice for social and political reform not merely because scientific evidence lends credibility to its political demands, but because breast cancer has been rearticulated as the experience of embodied social relations. This shared experience in turn necessitates collective action in the name of gender justice. As I noted in chapter 2, breast cancer activists were subject to attacks based not only on the science of environmental breast cancer but on their credibility as social and political actors. The resources used to discredit and/or silence activists reveals the powerful interests—both economic and political—that are threatened by their sustained struggle. Notably, BRCA carriers have never posed a united, collective threat to political and economic elites. This raises the question of biosociality, agency, and social change that I address in the next chapter.

6

Toward a Biosociality without Genes

This book opened with the case of "A.H.," a woman for whom a positive BRCA test made it thinkable to remove her breasts, ovaries, and uterus. The previous chapter ended with three vignettes of environmental social movement: civil rights and environmental justice, biomonitoring and chemical pollution, and breast cancer and gender justice. In all cases, the body figures centrally as that which both enables and compels action. Yet the bodies of these discourses are different in both the ontological and epistemological senses of the term. Embodied existence (which can also be understood as body practices) is the basis for forming connections with others. But how will those connections be forged? Will they be constituted through knowledge of shared genes? Or of shared experience of economic policies, social hierarchies, and institutional practices? These different relations in turn make possible particular formations of body politics. In the case of A.H. and cancer genomics more generally, the body itself is acted on; it is a genomic body that calls forth relations rooted in genetic identity. In contrast, social movements act on the body's behalf, changing the world it inhabits. In social movements, the shared experience of suffering, itself not dependent on prior somatic identity, is what instills racial and gender categories with meaning. These categories, therefore, remain in flux: contested and variable. The result is more possibilities for collective action in social, cultural, and political worlds.

I have described the body of genomics as epistemologically unique, allowing scientists to describe and document disease etiology as the product of impaired genes. Theories of disease etiology then determine what counts as common-sense intervention. Ontologically, the body of genomics is unique insofar as heredity structures its relationship to other bodies and to institutions. These epistemological and ontological developments have political implications, as I have shown throughout the book—namely, that hereditarianism serves particular sorts of institutional interests (such as those of molecular biology) and legitimates status quo social arrangements. Political and economic elites benefit from the transfer of pathology from policy and practice to nature and the body; the producers of genomics discourse benefit by creating biological explanations that ensure continued funding for practical, realpolitik solutions to disease, solutions that are expanding in scope.

The body, I have argued, is the material and discursive site for the recuperation of gender norms, biological race, and neoliberalism. Heredity, a scientific concept but also the cultural logic that I have called hereditarianism, is foundational to the success of this recuperative work. Genetic risk, I have maintained, is not something the scientist or physician discovers but is a rhetorical construction, one that articulates a particular configuration of relations between patients, institutions, and industrial capitalism. The practices that hereditarianism calls forth are confined to medicine and public health, laying claim to an apolitical and nonideological stance, under the protective rhetorical cover of pragmatism. Medicine aims simply to deal with the effects of physiological change in the body. When people get sick, they need to get better; when they are at risk, they need to be removed from the site of exposure. The integration of genomics into medicine has been a relatively seamless one, in large part because genomics strategically benefits from this pragmatist tradition by offering new ways to diagnose and treat disease. It even expands, conceptually, who can occupy the subject position of patient and what it means to be diseased; indeed, genetic risk is actionable in much the same way as a diagnosis of disease is (as Nikolas Rose put it, the patients of genomic medicine are "pre-symptomatically ill").[1] More and more people, it seems, will be subject to the medical gaze and its attendant practices, thus further validating medicine's implicit political and ideological investments in the status quo.

Although medicine and public health lay claim to the nonideological, they are no less political than the discourses that appear external to them. As this book has shown, scientific objects and medical practices are inextricably tied to the world outside the laboratory and the physician's office. Nature does not produce the normal and the pathological; social discourses (including that of science) do. Medicine and, to some extent, public health, mystify the ways in which we embody social relations, thereby tacitly endorsing their current configuration. Genes and the bodies they inhabit, moreover, come to the researcher and doctor already inscribed by the social. They are symbolic as much as they are material; indeed, how bodies "matter" is very much a product of discourse.[2] No object of scientific investigation or medical intervention comes to us ready-made but is, rather, the product of interest-driven discourses. Chapter 3, for example, showed how the ovary is an important site wherein hereditary fate and (obligatory) reproductive desire converge. Genomics thus benefits from the ovary's social and symbolic importance; but it also reaffirms the place of the ovary (and the women tied to it) in the social and economic order—the ovary becomes an extension of heredity, of its manifest materiality. It becomes, in reconfigured form, the source of a fixed gender identity. By linking together the oophorectomy crisis in gynecology, feminist and queer theory's challenge to the conflation of gender and sex, and genomics' creation of an exceptional case—the BRCA mutation carrier—chapter 3 showed the unique and perhaps unexpected way in which heredity performs important recuperative work beyond the geneticization of disease already well documented by critical studies of genomics.

This book also explored how the dynamic and evolving knowledge claims of genomics intersect with culturally coded and inscribed bodily facticities in the case of race. Chapter 4 documented how genomics' interest in the health disparities between white and black women is more than an elucidation of complex biological processes involving ancestry and its probable role in differential breast cancer etiology. It is, as I showed, an investment in the biological race concept that is, simultaneously, an investment in the implicit assumption that racism does not sufficiently explain these disparities. Race, according to genomics, is something one inherits but does not embody. The pathology of institutional racism is thereby hidden through the (pathologized) genomic body, which in turn is racialized through a series of rhetorical displacements. Contemporary genomics is thus recast in chapter 4 as a discourse very similar to eugenics and scientific racism, insofar as the diseased body is taken to be a product of nature rather than particular social environments. African American women are urged to take greater care of their personal health by submitting to genetic counseling and testing. This model of health disparities is quite different from, and perhaps comes at the cost of, due consideration of the radical implications of theories of health disparities around synchronic understandings of race—radical because they explicitly or implicitly call for significant social change.

Likewise in chapter 5, I considered how the gene-environment model helps normalize the practices of industrial capitalism by introducing the notion of susceptibility into public health discourse. Chemical pollution is implicitly assumed to be an expected, banal feature of modernity, which necessitates a pragmatic, realpolitik response: specifically, the identification of small subsets of "truly" at-risk individuals. Genes in this model are taken to be the only source of accurate, actionable knowledge. And it is actionable knowledge confined to practices of the body in the private sphere. These practices include therapies to counteract susceptibility as well as the strategic movement of bodies in occupational and ambient spaces in order to manage exposures.

This book thus marks a significant departure from other scholarly work that has considered various responses to the genomics revolution from the perspectives of ethics and social theory. Ethics has addressed the very real impact that genomics has had on medical and public health practice, and what this impact means for women's health. Social theorists have described changes beyond the fields of medicine and public health, considering instead how genomics has effected a profound shift in identity formation and social relations.

Charged with considering questions about human dignity and freedom, bioethics asks how we might reconcile the needs of women at risk for inherited breast and ovarian cancer—needs that are rhetorically constructed but no less real—with an awareness of the dangers that unregulated genetic testing poses. When evaluating the issues that arise in genetic counseling and testing, bioethicists therefore consider whether the test should be administered, given the person's chance of testing positive for a genetic mutation (informed consent figures importantly here), and if testing is warranted, what sort of protocol best protects the person's autonomy as she moves forward with the information the test reveals. For instance, before a woman submits to a genetic test, it is considered ethical to offer her comprehensive counseling. The counselor explains the framework for interpreting test results, names the interventions available to the woman if she tests positive for a BRCA or other gene mutation, and reviews the risks and benefits of those interventions. She is even counseled on the consequences of negative test results—some women feel tremendous guilt for having escaped the "fate" of other family members. In addition to autonomy, then, bioethics ensures that the basic principles of beneficence and nonmaleficence are upheld: that the test and subsequent interventions (surveillance, surgery) help promote a woman's health without exposing her to unnecessary danger.

The basic principles of bioethics—autonomy, beneficence, nonmalfeasance, and justice (which holds that no one should be denied access to medicine and good health)—limit the field of analysis to the interactions between a woman, her counselor, and her physician. Feminist bioethics scholars have criticized this limited focus because it essentially concedes that the material conditions that brought forth the interaction or exchange cannot or should not be objects of analysis.[3] Moreover, conventional bioethics' main principles do not attend to the complexity of people's lives. Informed consent, for example, rests on a limited notion of autonomy, one that presumes that a woman can—or should—be entirely free from outside influences. Beneficence cannot account for changing meanings of "health" and "disease" brought forth by the genomics revolution in medicine. Is it, for example, always a good idea to promote health if by doing so we expose women to dangerous therapeutic treatments based on risk alone? Feminists have also expanded the notion of justice to include gender justice—for example, by drawing attention to the specific ways in which medical issues affect the everyday lives of women.

My hope is that the analysis in this book will help answer many of the questions raised by both conventional and feminist bioethics: Do genetic testing and subsequent interventions like prophylactic surgery uphold the principles of nonmaleficence and beneficence? Have alternatives to genetic testing been pursued? Is genetic testing, as a technology, the product of particular sets of social relations, with its design and use overdetermined by values attached to those social relations? On what grounds do women resist, and what can we learn from their resistance?[4] How can we problematize autonomy and choice and the subject positions they require? More important, in what ways is genomics discourse constitutive of the subject positions that women come to inhabit? How does the relationship between knowledge claims and interventions become intelligible? And how might we understand these interventions beyond the limited framework of the doctor-patient relationship? Addressing such questions requires an analysis of the cultural and social norms that are called on in genomics discourse to fashion medical subjects; how these subject positions serve particular kinds of recuperative, ideological work (especially concerning gender and race); and how interventions such as surgery, explanations of health disparities, and public health surveillance are both evidence of changes in biomedicine that affect women as patients and normative statements about the world in which they live.

But even feminist bioethics must, for the most part, work within the terms set by hereditarianism and biomedicalization more broadly.[5] The subjects of bioethical discourse—BRCA carriers—are defined by their genome. Their agency is therefore an outgrowth of that identity. When BRCA carriers make decisions based on the principles of either conventional or feminist bioethics, they are acting as mutation carriers. Feminist bioethicists, in turn, are acting on their behalf. More to the point, feminist bioethicists are acting on behalf of what Rose has termed a "biosocial community," one that would not exist if its members had not submitted to a test and whose identity is now structured (even if just partially) around the knowledge the test produced. Feminist bioethics must, therefore, tailor its analysis to what counts as ethical and just action for people who are so defined. The ethical discourse it produces is thus limited to those women who submit to testing or who otherwise internalize the logic of hereditarianism. It does not address, for instance, what satisfies the conditions for ethical life for the women who resist or otherwise fall outside the scope and practice of genomic medicine.[6]

For those women who inhabit the subject position of BRCA mutation carrier, bioethics discourse and the hereditarianism it indirectly concedes make possible a particular kind of subjectivity and relationality—BRCA carriers, for example, are connected to each other by virtue of their genetic identity and its attendant risk. Breast cancer and breast cancer risk are, within this frame, the products of evolutionary history that leave an indelible mark on their hosts. Neither laws nor norms, other than those dealing with genetic privacy and discrimination (which limits "rights" discourse to a "negative" rights frame), need to be altered on behalf of the BRCA carrier. This state of affairs is in stark contrast to the social movement activity around breast cancer that I describe in chapter 5. Heredity as an organizing principle of bioethical discourse thus limits the possibilities for theorizing and enacting political and social change.

Or does it? Biosociality, according to Paul Rabinow, is a particular type of relationality made possible by the HGP and the profound collective and cultural changes it augured. The genome, or simply heredity, has become the defining feature of social life: disease, behavior, identity itself are all inextricably bound up with genes and their variants. For Rabinow, it is possible—although by no means guaranteed—that biosociality can be a productive type of association, one that at the very least is much more complicated and expansive than the interactions women have with their genetic counselors and physicians. It is an association that is radically undetermined: biosociality has no inherent politics; according to Rose, biosociality is merely a social structure for the enactment of particular projects, such as the demand for research dollars and other resources to investigate genetic disorders. And sociologists have shown how disease communities can often expand the field of political action considerably.[7]

In a somewhat similar vein, Margrit Shildrick has called on feminists to recognize and embrace new possibilities of identity and connection brought on by bioscience and, in particular, genetics. Shildrick argues that explorations of the human genome have called into question many of our most sacred truths. Although institutions often succeed in recuperating the normative body (what Shildrick describes as an imposing of order on the disorder brought about by bioscience), they cannot entirely contain the new relational possibilities that emerge in the wake of genomics' conceptual obliteration of essential identities. What it means to embody genomics will always exceed or be underdetermined by ideology. Shildrick notes that genomics has routinely destabilized subjectivity and identity; the genome, as it turns out, does not provide the basis for firm boundaries between human groups or between humans and nonhumans. Genomics forces us to continually question and negotiate what it means to be human, opening up new possibilities for the enactment of world-transforming kinship. She writes:

> The very real ethical questions posed by bioscience in general and genetics in particular arise in the context of both an attempted shutting down of disorderly options, and a proliferation of uncontainable transformations and unforeseen connections. Where genetics may always threaten new patterns of oppression, we might also see new and positive possibilities of fluidity and connection. In the face of such a dynamic, the ethical undertaking cannot be to impose the order of answers that will prevail unchallenged over time, but rather to abandon the desire for the solid ground where certain analytic categories and concepts are fixed in advance. In consequence, the search for resolution, even for settled attributions of right and wrong, is no longer the primary concern. The task, instead, is to refuse an illusory closure and to continually reopen the questions themselves.[8]

In this passage, Shildrick situates herself within a group of scholars attempting the very important work of rethinking the terms of ethical principles without recourse to the very categories they are trying to destabilize.

The notion of biosocial communities, for example, helps us rethink, in important ways, the autonomous subject of bioethics, a subject that has never really existed and that effects a dangerous obfuscation of the ways in which genetic subjects are culturally inscribed. Certainly, members of these newly constituted biosocial communities have engaged—and will continue to engage—in a range of practices that cannot be anticipated in advance but that nonetheless have the potential to shape ethical guidelines and public policy.

Much of what I have argued in this book is not altogether incompatible with an antifoundational reading of the body and its emphasis on becoming over being and on openness to contingent, identity-based politics. Nevertheless, it makes little theoretical or practical sense to depend on an artifact of being (genes) for the promise of becoming. Genes are fixed properties of the biological body; even Marxist biologists and epigeneticists admit there is not much one can do with a defective gene inherited at birth. And although biological development is complex and varied, there are limitations to what the organism can become. More to the point for the issue of genes and biosociality, the latter is the outgrowth not of uncertainty and complexity, but of fixed genetic identity. Indeed, biosocial practices are only possible because of that identity—BRCA carriers are a biosocial community (as shown, for example, by collective and individual blogs) because they all received a positive genetic test result. And if, as Rose contends, we understand ourselves as ethical beings through practices of the body, if that body is a genetic body, ethical life will be imagined accordingly.

This, to my mind, is the central problem with the notion of biosociality. As I argued above with regard to bioethics, biosociality limits the kinds of alliances—and, by extension, politics—that are possible. More important, the actions of biosocial communities are concerned principally, even exclusively, with disease, and a biomedicalized version of disease at that.

As I argued in the book's introduction, the social change and reforms of the last hundred years or so mean that hereditariansim can no longer be imposed—it must be internalized. It is entirely possible, I want to suggest here, that biosociality enables this internalization by embracing what Rose calls the "somatic self":

We are increasingly coming to relate to ourselves as "somatic" individuals, that is to say, as beings whose individuality is, in part at least, grounded within our fleshly, corporeal existence, and who experience, articulate, judge, and act upon ourselves in part in the language of bio-medicine. From official discourses of health promotion through nar-ratives of the experience of disease and suffering in the mass media, to popular discourses on dieting and exercise, we see an increasing stress on personal reconstruction through acting on the body in the name of a fitness that is simultaneously corporeal and psychological. Exercise, diet, vitamins, tattoos, body piercing, drugs, cosmetic sur-gery, gender reassignment, organ transplantation: the corporeal exis-tence and vitality of the self has become the privileged site of experi-ments with the self.[9]

These bodily practices are, arguably, available only to those with the privilege to exercise them. This raises the questions, then, of how and whether these practices of the self also constitute resistance to ideologies of health. Might practices of the self, so construed, also be mechanisms for social control and for imposing limits on membership in the polity?

Let me return to the example of race to draw out this particular point in greater detail. Bioethics principles can help guide the administration of genetic tests, and critical approaches can add to this by considering the ways in which race, gender, and the intersection of the two must in-form the application of these principles. For example, bioethics addresses whether tests are available to black women, whether genetic counseling is duly attentive and sensitive to their information and decision-making needs, and whether counselors and physicians are more or less likely to recommend radical interventions like prophylactic surgery.

However, because bioethics must limit its framework of analysis to the doctor-physician relationship—even if it interrogates the material con-ditions that give rise to and structure this relationship—it must act on behalf of black women as a group of women defined in some way by their genes. It cannot, then, question the ethicality of the very creation of the racialized testing subject, and it cannot ask whether the social des-ignation "African American" should justify an analysis of her genome. Indeed, it is the racialized testing subject itself that makes the bioethi-cal query possible. And the principle of justice can act as a further im-pediment to exploring the ethicality of racial medicine. Because BRCA testing, for example, is readily available to middle-class white women, bioethicists are obliged to problematize the low rates of testing among African American women in the United States. Within the terms of dis-cussion set by traditional or conventional bioethics, justice would be served by aggressive outreach in the African American community in order to ensure more uptake of the BRCA test (as I suggest in chapter 4, there are good reasons why black women resist being tested). Even feminist bioethics—by expanding the notion of justice to consider, for example, why breast cancer risk is increasingly being defined in genomic terms for black and white women alike—would not have the conceptual means to question the very existence of the "black" risk subject.[10] The very production of bioethical discourse is dependent on the existence of this subject position.

Biosociality theorists like Rose see nothing inherently dangerous in the practice of identifying black women as potential recipients of genetic testing. A biosocial community, even if race-based, is nothing more than the expression of the hopes and aspirations of a community called forth by shared disease heritage. Maren Klawiter, in a related vein, has termed these collectivities "disease regimes." By way of contrast, consider the argument of Troy Duster about sickle cell as both a disease and a social phenomenon. According to Duster, sickle cell has been labeled a "black" disease, largely because the meaning of disease genes is overdetermined by the racial designation of the bodies they inhabit. African American physicians, researchers, and communities rallied, demanding more research and medical attention to African Americans with sickle cell anemia as well as those with sickle cell trait in need of genetic counseling. The result, Duster documents, was the routine testing of African Americans, even when unwanted and unnecessary, as well as widespread discrimination in employment and health for those who were only carriers of the recessive gene for sickle cell anemia.[11]

My interpretation of Duster's pathbreaking study is that race overdetermined the meaning of the sickle cell gene, thereby demonstrating that biosociality is never autonomous from institutions and social structures of racism more broadly. In fact, biosociality may, more often than not, be complicit in their operation. Duster's study draws attention to the failure of gene-inflected biosociality (and, perhaps, disease regimes in general) to address some of the more fundamental problems facing African Americans with sickle cell anemia: access to health care, economic security, and the considerable resources needed to manage the disease's symptoms. Extending this analysis to the case of genomics and health disparities, Jacqueline Stevens argues that the advocacy of high-profile African American geneticists has obscured genomics' inherent racialism and eventual reification of race. And, she points out, if the financial incentive for integrating genomics into health disparities research did not exist, neither would its support in the African American research community.[12]

The example of social movement that I described at the end of chapter 5 shows how the body can be the center of different epistemologies and politics. Environmental justice activists, for example, do not claim that race explains why they get sick from chemical pollution; it explains, rather, why the pollution is there in the first place. Industrial capitalism and institutional racism—their historical shifts, their specific practices—together give meaning to "race." As a term, it is meaningful only by way of a synchronic analysis of racist practices. These racist practices are possible because the notion of racial inferiority remains widespread—a phenomenon for which genomics is partly responsible. And even when a social movement is organized around a specific disease—as in the case of breast cancer or sickle cell anemia—there are, I think, fundamental differences between a disease community and a mutation community. To identify with another based on a disease diagnosis initially limits the range of actions to that disease. But in many circumstances, there are opportunities for expanding the meaning of that connection. Breast cancer activism eventually led to environmental breast cancer activism, opening up the possibility for women to consider their position not only within a biomedicalized disease regime but within social relations more broadly. Environmental breast cancer activists identify themselves both as members of a disease community—all sharing a diagnosis of breast cancer—and as gendered, raced, or classed subjects with a particular relationship to industrial capitalism and the political entities charged with containing its destruction. The subject identity connected to a disease community is, as a consequence, fluid and contingent. In contrast, the disease community comprising BRCA carriers does not permit similar fluidity in identity and connection. The mutation fixes one's identity; it is not something that can be altered. It is an identity rooted in evolutionary history, not social history. Social movements thus politicize disease in a way that gene-based biosocial communities do not.

In environmental social movement, the diseased body and the body-at-risk are together evidence of the materiality of social relations. But importantly, the body is not the locus of action, as in the body practices of somatic ethics. The locus of action is, rather, at the level of social and economic policy. In this way, social movements directly challenge genomics and its investment in an atomistic worldview. The atomized society of industrial capitalism, says Richard Lewontin, is matched in genomics by a "new view of nature, the reductionist view." The ideology of modern science, he says, including modern biology, "makes the atom or individual the causal source of all the properties of larger collections."[13] For biosociality to challenge this worldview, it can no longer be the outgrowth of indelible, biologistic, bodily attributes and their implicit valorization of atomistic identity and politics.

What is required, I believe, is a notion of biosociality without genes, perhaps even without biology, as exemplified by the actions of civil rights, feminist, and environmental health activists. Whereas the biosociality that social theorists describe is animated by the disclosure of one's genetic mark, biosociality in this other sense is animated by the potential to act on one's experience, in connection with others.[14]

NOTES

PREFACE

1. Treichler 1999, 2 (italics in the original).

CHAPTER 1

1. Spear et al. 1999, 552.
2. "BRCA" stands for "breast cancer," although mutations of BRCA1 and BRCA2 can increase risk for ovarian and other cancers as well. BRCA research is an early example of a research program known as genomics, which I define broadly as the investigation of the role of genes in health and disease based on knowledge of the human genome—in particular, on the characteristics of the genome of an individual or group.
3. With regard to AIDS, Treichler explains: "The AIDS epidemic is cultural and linguistic as well as biological and biomedical. To understand the epidemic's history, address its future, and learn its lessons, we must take this assertion seriously. Moreover, it is the careful examination of language and culture that enables us, as members of intersecting social constellations, to think carefully about ideas in the midst of a crisis: to use our intelligence and critical faculties to consider theoretical problems, develop policy, and articulate long-term social needs even as we acknowledge the urgency of the AIDS crisis and try to satisfy its relentless demand for immediate action" (1999, 2). See also Levins 2000.
4. Hayles 1993.
5. Waldby 1996, 5.
6. Harding 2006, chapter 7.
7. Ibid., 4–5. See also Harding's distinction between "bad science" and "science as usual" (Harding 1991, chapter 3). According to Rouse (2004), these insights are what set feminist science studies apart from studies of the sociology of scientific knowledge.
8. Jasanoff 2004, 3.
9. See Longino 1990.
10. One could say it is a version of the idea that medicine functions as a state ideological apparatus.

11. See Stabile 1995.
12. Centers for Population Health and Health Disparities 2007, 15. Krieger's corpus of scholarship has consistently explored this point. See, for example, Krieger 2001 and 2005. Also see Levins 2000.
13. Levins 2000, 15.
14. See Brooks and King 2008.
15. Stevens (2008) has introduced the terms "diachronic" and "synchronic" to describe differing approaches to employing the race concept in biomedical research, which I have found enormously helpful.
16. Stevens 2008.
17. Delgado and Stefanic 2001, 15. See also Delgado and Stefanic 2000.
18. Keller 1992b; Hacking 1983.
19. Jasanoff 2004, 2.
20. Krieger's work exemplifies this progressive turn. For critical approaches to "health," see, for example, Metzl and Kirkland 2010.
21. There is, of course, the "gene-environment" model espoused by epidemiologists and geneticists in which neither "gene" nor "environment" ought to take precedence over the other, and in which one cannot be reduced to the other. I critically engage this model throughout the book, especially chapter 5.
22. E. Martin refers to the discourse of reproductive biology to make this point: "Bodies are organized around principles of centralized control and factory-based production. Men continuously produce wonderfully astonishing quantities of highly valued sperm, women produce eggs and babies (though neither efficiently) and, when they are not doing this, either produce scrap (menstruation) or undergo a complete breakdown of central control (menopause). The models that confer order are hierarchical pyramids with the brain firmly located at the top and the other organs ranged below. The body's products all flow out over the edge of the body, through one orifice or another, into the outside world. Steady, regular output is prized above all, preferably over the entire life span, as exemplified by the production of sperm" (1990, 121–22). For her work on "flexibility," see E. Martin 1994.
23. Haraway 1989.
24. Cooper 2008.
25. E. Martin's pathbreaking study *The Woman in the Body* (1987) revealed that class position influences the degree to which women internalize, and put into practice, the assumptions of dominant medical paradigms. According to her ethnographic research, middle-class women are more likely to describe menstruation in medical or clinical terms, working-class women in experiential or phenomenological terms. See Sunder Rajan 2006 for an excellent analysis that connects the discourse of genomics to particular concepts and practices of capitalism, such as surplus value.

26. Lowe 1995. In this way, Lowe departs from the work of biosociality theorists like Rose (2007). There are, I believe, two problematic assumptions on which the biosociality concept rests: (1) that the somatic self is more or less a new development (when in fact women and people of color have long been forced to lead corporeal existences); and (2) that the somatic self can be deciphered and theorized independent of economic structure. I do not disagree with biosocial theorists like Rose that developments in biomedicine and in the culture at large have brought about a particular somatic subjectivity, only with whether this constitutes a break with past cultural logics. I explore this point more fully in chapter 6.
27. This is another point of departure for Lowe (1995): he calls on the critic both to account for economic structure and to attend to the governing practices of particular systems of oppression that cannot be reduced to it.
28. Hartmann 1997; Barrett 1997; Nicholson 1997.
29. Lowe 1995, 109.
30. Sunder Rajan 2006; Cooper 2008; Waldby and Cooper 2008.
31. Rose 2006, 160–61.
32. See Feenberg 1991 for a critical theory of technology perspective.
33. Stevens 2008.
34. Stabile 1992.
35. Stabile 1995, 415.
36. This is one way in which I part company with scholars like Rose who draw on Foucault to make the case for interpretations of health discourses that disavow any notion that biomedicine acts as an ideological state apparatus. My approach here is that Foucault is helpful in understanding the production of new subjectivities (in particular how work on the body replaces explicit ideological appeals), but that new subjectivities do not necessarily tell us about the working of dominant interests. I say more about this in the conclusion.
37. Oyama, Griffeths, and Gray, 2001.
38. See Nelkin and Lindee 1995 and van Dijck 1998 for an appraisal of popular representations of genes.
39. I provide a more complicated reading of genomics in chapter 5, especially with regard to gene-environment interaction.
40. Haraway 1989, 15.
41. Happe 2000.
42. Ohmann 1996.
43. Butler 1990.
44. Harding 2006, 82. See also Condit 1999a.
45. This perhaps explains the somewhat surpring results of ethnographic work that has studied the attempt of scientists to disavow race in favor of other terms, employing different software programs to do so. See, for example, Fujimura and Rajagopalan 2011.

46. Barad 2003.
47. Jasanoff 2004, 5.
48. My theoretical understanding of scientific discourse is very much indebted to the work of the European historian of science Georges Canguilhem. See, for example, his *The Normal and the Pathological* (1989). See also Foucault (1994).
49. Stabile 1992, 180.
50. For feminist scholarship on the status of the fetus, see Morgan and Michaels 1999.
51. At least, for the most part. As this book went to press, the increase in public discourse in which white single women's use of birth control was attacked was quite striking.
52. See, for example, the forum "Is Race Real?" (2005).
53. Fields 1990, 114.
54. M. Brown, Carnoy, Currie, and Duster 2003, 223–24. Like Brown and coauthors, I use the terms "African American" and "black" interchangeably, both for stylistic reasons and because racial discrimination is experienced by people who do and do not identify as African American.
55. Ibid., 2003, xi.
56. Ibid., 226. See also Collins 2004.
57. Gao et al. 1997.
58. Stevens 2002; Reardon 2004.

CHAPTER 2
1. Keller 1992a.
2. A press conference announcing the project's (near) completion was conveniently timed to occur in the year 2000, allowing advocates to frame the sequencing of the human genome as yet another marker of the new millennium. Then, in April of 2003, it was announced that the sequence was *truly* complete, this time coinciding with the fiftieth anniversary of James Watson and Francis Crick's seminal publication on the structure of DNA.
3. Kevles 1985. For additional histories of eugenics, see Mazumdar 1992; Kline 2001; Ordover 2003.
4. The first family study, *The Jukes*, was published in the late nineteenth century by Richard Dugdale. In the book, Dugdale documents the effects of heredity by cataloging the number of criminals, alcoholics, prostitutes, and the like within the Jukes family. This study would be cited by eugenicists for some time as evidence that pathological behavior is inherited. Nicole Rafter notes: "Eugenics had enormous impact on the direction taken by the newly developing disciplines and professions of criminology and criminal justice, psychology and psychometry, sociology and social work, and statistics. Through legislation it shaped social policy governing crime control, education, liquor consumption, marriage and birth control, mental retardation,

poor relief, and sterilization—all topics with which the family studies deal"
(1988, 30). The discovery of Mendelism in 1900 appeared to provide solid
scientific evidence of the mechanism by which traits are passed down among
generations, although researchers did not know exactly what the substance
of heredity was. The concept of the gene would be hypothesized for the first
time by Wilhelm Johannsen in 1909. The following excerpt from a 1913 issue of
Cosmopolitan captures this enthusiasm: "The long controversy about the relative influence of heredity and environment has been settled for all time. We
know now that the possibilities of any individual are determined before birth"
(Goodhue 1913, 148).

5. The laboratory opened in 1904 with funds from the Carnegie Institute in
    Washington, D.C., for the study of evolution.

6. Between 1910 and 1918, Harriman gave half a million dollars to the ERO. John
    D. Rockefeller contributed an additional $20,000 a year for four years.

7. Before the American Eugenics Society was founded, activity included the
    formation of local eugenics groups across the country: the Galton Society,
    which met at the American Museum of Natural History (New York); the Race
    Betterment Foundation, in Battle Creek, Michigan; and "eugenics education"
    societies in Chicago, St. Louis, Wisconsin, Minnesota, Utah, and California.

8. See Kevles 1985, chapter 3.

9. For a history of the progressive era, see Danbom 1987.

10. Haller 1984, 5.

11. Southern and Eastern Europeans were considered nonwhite and, thus, infe-
    rior. They were also blamed for widespread labor unrest and some of the more
    radical and militant labor activism.

12. For a social history of the 1920s, see Dumenil 1995.

13. The term "germ plasm" was a precursor of the word "gene."

14. Wiggam 1922, 647.

15. Wiggam 1926, 25.

16. Between 1907 and 1963, almost thirty years after the demise of the eugenics
    movement, thirty states had sterilization programs and over 60,000 persons
    were forced to undergo this surgery (Reilly 1987, 161). As of 1987 sterilization
    laws were still on the books in 23 states (ibid., 166).

17. Kevles 1985.

18. Ibid., 56.

19. See Happe 2000.

20. See Ohmann 1996 for a detailed look at hegemony theory and the professional
    and managerial class. See Happe 2000 for the application of Ohmann's theory
    to the specific case of eugenics discourse in mass-circulation magazines of the
    early twentieth century.

21. Frank 1922, 317.

22. Wiley 1919, 167.

23. Quoted in Kevles 1985, 320, note 23.

24. Quoted in Gould 1996, 212.

25. Higham 1988, 191.

26. This is the word used by historians to describe the eugenics movement that preceded "reform" eugenics.

27. Mainline eugenicists held that "race suicide" was the likely consequence of uncontrolled breeding of the unfit and the failure of the superior classes (that is, whites in the middle and upper middle class) to reproduce.

28. Paul 1995, 116–17.

29. Kevles 1985, 169.

30. Ibid., 173.

31. Ibid., 175. As I show in chapter 4, the word "race" has, for the most part, been replaced by "population," yet many geneticists, implicitly or explicitly, assume that populations look much like races.

32. Muller was also known as a one-time leftist who moved to the Soviet Union in order to enact his vision of a socialist eugenics. According to Paul, in 1934 Muller sent a copy of his eugenicist manuscript to Joseph Stalin, "accompanied by a letter effusively praising Bolshevism and excoriating the race and class bias of capitalist eugenics. Muller's manuscript described a vast program for the artificial insemination of women with sperm of men superior in intelligence, talent, and social feeling" (1995, 118).

33. Ibid., 120.

34. Kevles 1985, 196–97.

35. Ludmerer 1972, 168.

36. Kevles 1985, 198.

37. Ibid., 253.

38. Biochemical tests included urine tests that could diagnose metabolic disorders (such as phenylketonuria), some of which were thought to be Mendelian traits.

39. Kevles 1985.

40. According to Kevles, demand for genetic counseling was low in the 1940s and 1950s, but this changed considerably after 1960. In 1960 there were between thirty and forty clinics; by 1974 the number had reached nearly 400 (1985, 257).

41. See Kay 1993. An example of such a new program was the one started at the California Institute of Technology. Molecular biologists included scientists like Morgan, the reform eugenicist.

42. Paul 1995, 115.

43. Kevles 1985, 206.

44. The "unit character" concept held that single genes were linked to single phenotypic outcomes.

45. Kay 1993, 37.

46. Mazumdar (1992) argues that the term "reform eugenics" is really a misnomer,

since there was considerable interest, in the United Kingdom, both in traits such as feeble-mindedness and psychosis as single-gene recessive traits and in sterilization for preventing their transmission. In the United States, she notes, eugenic sterilizations continued beyond the 1930s. For a history of sterilization in the United States, see Reilly 1987.

47. Kay 1993, 8-9.
48. Ibid., 9.
49. Ibid., 44.
50. Ibid., 47.
51. Quoted in ibid., 36.
52. Ibid., 45.
53. Olby 1974; 1990, 506.
54. Olby 1990, 513–14.
55. Watson and Crick 1953b. For a short biography of Rosalind Franklin, see Hubbard 2003. According to Hubbard's research, Watson and Crick could have never put together the model of DNA had they not had access (unbeknown to her) to Franklin's X-ray diffraction images of DNA.
56. Watson and Crick 1953a, 966. For a social history of the "code" paradigm, see Kay 2000.
57. Watson and Crick 1953a, 966.
58. Kuhn 1970.
59. Rabinow 1992.
60. Keller 1992b.
61. Kevles 1985, 209.
62. This is according to Paul 1995. Kevles (1985) dates its founding to 1950.
63. Kevles 1985, 223.
64. Ibid., 231.
65. Ludmerer 1972; Kevles 1985.
66. Ludmerer 1972.
67. Kevles 1985, 254.
68. Neel held a joint appointment in the Medical School and the Laboratory of Vertebrate Biology at the University of Michigan beginning in the late 1940s.
69. Pregnant women participated in the study in order to get special provisions during and after the US occupation of Japan. Midwives were compensated with money, and "it was also stressed to them what a unique contribution to the scientific knowledge of the entire world they were in a position to make" (Beatty 1991, 302). Research still continues under the Radiation Effects Research Foundation, a research collaboration between the United States and Japan.
70. "Race of monsters seen" 1952.
71. "Geneticist warns on radiation rise" 1954, 117.
72. Ibid.
73. Quoted in ibid.

74. Quoted in ibid.
75. Leviero 1956.
76. Ibid.
77. Kaempffert 1956.
78. Trumbull 1955, 6. See also Holmes 1955.
79. Muller 1955. See also Blair 1955. See note 32 for more information regarding Muller's earlier leftism.
80. See Keller 1992b.
81. Watson and Crick's fondness for describing the object of their work as "life" itself is well documented. The word appeared in Watson's research notes (Olby 1974), and, according to Watson, it was Crick's reading of Schrodinger's *What Is Life?* that sparked his interest in DNA. Watson also called DNA the "Rosetta Stone," the "true secret of life," and the "most golden of all molecules" (Watson 1968, 18–21).
82. "The Secret of Life" 1958, 50. See also Platt 1962.
83. Engel 1962, 39.
84. Osmundsen 1962.
85. Ibid. Osmundsen is not referring to Watson and Crick's theory regarding the structure and function of DNA. Rather, he is referring to speculation that scientists might figure out the sequence of the base pairs.
86. Wilkins was credited with taking "pictures" of DNA molecules using electron microscopes.
87. W. Laurence 1962.
88. Crick 1958.
89. The popular understanding of DNA is that its chemical bases provide the instructions for protein development and that RNA is merely a copy or messenger of the DNA code. However, recent research suggests that RNA is older than DNA and much more instrumental in guiding the development of proteins after it leaves the cell's nucleus.
90. Crick 1970, 562. Certain RNA viruses are called "retroviruses" because they contain the enzyme "reverse transcriptase," allowing RNA in the virus to produce DNA.
91. Ibid., 563.
92. Olby 1990, 510.
93. Levins and Lewontin 1985, 180. Auguste Weissman was an early-twentieth-century researcher who, responding to the theory of acquired characteristics, argued that the human germ plasm remained unaltered during an individual's lifetime.
94. The theories of Lamarck informed the bulk of genetics research in the Soviet Union, especially under the stewardship of Lysenko. See Levins and Lewontin 1985; Graham 1993. Levins and Lewontin, although sympathetic with the more nuanced explanation of biological development that Lamarck's work afforded,

nevertheless criticize Lysenko for the dogmatic, (and, as it turns out, anti-progressive) appropriation of Lamarck for Soviet plant genetics and political economy more generally.

95. Levins and Lewontin 1985, 168–69.
96. For this, I rely principally on Chafe's (1999) history of the period.
97. I define this in the section on eugenics above.
98. Chafe 1999, 138.
99. Ibid., 144.
100. Keller 1992a, 288–89.
101. Ibid., 289.
102. Lewontin 1991.
103. Ibid., 67.
104. Ibid., 64.
105. The HGP included sequencing the human genome, mapping genes, and sequencing the genomes of other organisms such as bacteria (*Escherichia coli*), yeast (*Saccharomyces cerevisiae*), fruit flies (*Drosophila melanogaster*), and worms (*Caenorhabditis elegans*). The project's goals also included the development of research technologies (such as sequencing technologies), technology transfer, and programs for publishing and distributing research data.
106. Wade 2003. The original estimate was $3 billion.
107. For book-length histories, see Bishop and Waldholtz 1990; J. Davis 1990; Wingerson 1991; Lee 1991; R. Shapiro 1991; Wills 1991; Cook-Deegan 1994.
108. Mapping simply refers to locating a gene on a chromosome; sequencing is the word used to describe identifying the chemical make-up of the gene.
109. Kevles 1992, 18–19.
110. Kevles 1997, 274.
111. Kevles 1992, 19.
112. Kevles 1997, 272.
113. Quoted in Kevles 1992, 22.
114. Quoted in Gilbert 1987, 27, 33.
115. Kevles 1997, 275.
116. Ibid., 275–76.
117. Ibid., 277.
118. Kevles 1992, 1997; Cook-Deegan 1994; Fortun 1995.
119. Although both the NIH and DOE declared a shared interest in both sequencing and mapping, the DOE was principally in charge of developing new sequencing and analyzing techniques, using several of the national laboratories and working with members of the computer science and physics communities. The NIH was principally responsible for mapping and for creating public databases, and it had sole responsibility for mapping model organisms. Additionally, the NIH took principal responsibility for producing and circulating medical applications of genomic research.

120. For a discussion of this aspect of the HGP debate, see Mitchell and Happe 2001.

121. David Baltimore (1987), a Nobel laureate and the head of the Whitehead Institute for Biomedical Research at the Massachusetts Institute of Technology, expressed this common concern. Some scientists went even further and argued that even if no direct trade-off occurred, it was still the case that any new money should go to the general coffers of the NIH in order to ease "the strained circumstances of basic biomedical research" (quoted in Kevles 1997, 287).

122. Many pointed out that much of the genome comprises "exons" (noncoding regions also known as "junk DNA") and that sequencing them would be useless. Many also argued that sequencing the introns (coding regions) would not necessarily advance biomedical science (see, for example, Weinberg 1987; Ayala 1987). Of course, this argument was dependent on the state of knowledge at the time; "junk" DNA is no longer considered so.

123. David Botstein, involved in the development of mapping technology, argued at the Cold Spring Harbor meeting that a genome project threatened to render biologists "indentured servants" to "big science"—a widely shared sentiment among scientists who viewed the proposed HGP as antithetical to the extant NIH funding infrastructure that encouraged competition and diversity of research (quoted in Kevles, 1997, 276). Many biologists thought that HGP-related work—which undoubtedly would determine the careers of countless future researchers—would be "tedious, routinized, and intellectually unrewarding" (ibid., 277).

124. Fortun points out that the National Research Council committee was comprised mostly of the key, "core" proponents of the HGP, so it was not particularly "objective" (1995, 420).

125. Kevles 1997, 278. Physical mapping entails finding particular sets of chemical bases that serve as spatial markers on the genome but that do not hold any particular biomedical value. Genetic mapping entails finding the location of genes thought to be responsible for disease.

126. Kevles 1992, 26.

127. Opposition nevertheless remained. In 1990 the biochemist Martin Rechsteiner circulated a "Dear Colleague" letter advocating that the HGP be halted, in large part because, according to him, any benefit to biomedical science would not justify the cost (see Kevles 1997; Fortun 1995, 368, 373). Rechsteiner claimed that the HGP represented "mediocre science [and] terrible science policy" (quoted in Fortun 1995, 371). Also in 1990, a letter-writing campaign was launched by another biochemist, Michael Syvanen, who advanced the same arguments as Rechsteiner (ibid., 367). And later that year, Bernard Davis and almost all of his colleagues in the Department of Microbiology and Molecular Genetics at Harvard Medical School published a letter to the

editor in *Science* urging reconsideration of the project (Kevles 1997, 286). In 1990 Rechsteiner and Davis, two of the more vocal critics, were both asked to testify at a Senate hearing convened by Domenici. Yet only one of the two actually testified, and by then eleven of the twelve journalists who had attended the hearing had already left: "It was indeed evident that Domenici (nor for that matter any of the other senators who had shown up the hearing that day) was not particularly interested in hearing and considering criticism of what had been a favored project of his for several years" (Fortun 1995, 390). But even in 1991, Wyngaarden conceded that "if you took a vote in the biological sciences on the project, it would lose overwhelmingly" (quoted in Kevles 1997, 286).

128. Restriction enzymes allow geneticists to "cut" sections of DNA from one cell and bind it to the DNA of another.

129. Krimsky 1982. See also S. Wright 1994.

130. Interestingly, according to the media studies scholar Rae Goodell (1986), the rDNA controversy was short-lived; it did not take long for leaders of the scientific community to close ranks and support unrestricted research in this area. This was largely accomplished by the demonization of critics in public by other scientists (in an about-face from previous willingness to debate the issue of regulation), embargoes on journalists attending genetics conferences that prevented them from consulting outside experts, and some researchers refusing to grant journalists interviews unless they promised not to quote critics in the same article. For an excellent rhetorical analysis of the hearings in Cambridge, Massachusetts on a proposed ban on rDNA research, see Waddell 1990.

131. R. Wright 1990. According to Fortun (1995, 415–17), Watson had to have known that Congress would demand some sort of ELSI program. Kevles notes: "Watson was not only undaunted in his commitment to ethics but also, it would appear, shrewd. His policy undoubtedly helped defuse anxieties about the prospect of a genome project indifferent to or unrestrained by ethical considerations" (1992, 35). As of 1999, of the 172 studies granted ELSI funds, 73 (42 percent) were on topics related to educating the public and health professionals, assessing attitudes about genetics, and counseling individuals facing the choice of undergoing genetic testing or those who, having already undergone tests, have to negotiate the psychological, cultural, and interpersonal significance of the test results. See Juengst 1996 for a criticism of the ELSI program and the overrepresentation of these topics.

132. Koshland 1989, 189.

133. Office of Technology Assessment 1988, 85.

134. Chafe 1999, 473.

135. Ibid.

136. Ibid., 474.

137. The Violence Initiative, a federal project enthusiastically supported (despite criticisms of some of its advocates' public statements) purported to explain the genetics of violence by studying young black and Latino subjects. It was canceled before receiving any funding, although as Sellers-Diamond (1994) points out, what was canceled was the initiative (a coordinating effort), not violence and race research per se. See also R. Wright 1995. For a more general treatment of the ideological and historical dimensions of violence research, see Lewontin, Rose, and Kamin 1985; Rafter 2008.

138. Genetics is not, technically, synonymous with genomics, even though I use both in this book to denote what might be termed the "science of heredity." "Genomics" is typically taken to be the contemporary, post-HGP word for heredity research, while "genetics" is a more general term denoting the investigation of both hereditary and adult-onset mutations.

139. Sunder Rajan 2006.

140. Sunder Rajan explains: "There is biological information, and the biological material (cell or tissue) from which the information is derived, material that subsequently becomes the substrate of experiments that validate the leads suggested by the information. In the process, information is detached from its biological material originator to the extent that it does have a separate social life, but the 'knowledge' provided by the information is constantly relating back to the material biological sample. The database plays a key intermediary role in the translation of 'information' to 'knowledge': in this case specifically knowledge that is relevant to therapy. It is knowledge that is always relating back to the biological material that is the source of the information; but it is also knowledge that can only be obtained, in the first place, through extracting information from the biological material. The abstraction of information away from the biological material has a specific function in making therapeutically relevant knowledge. This is also why it is so easy to intuitively conceptualize the generation of information as 'inventive,' and therefore as something that can legitimately be owned" (ibid., 42).

141. Keller 1992a, 295.

142. Lewontin 1991, 7.

143. Keller 2000, 146.

144. Ibid.

145. Commoner 2002, 43, 42. Many of these cellular process have been known for some time. Splicesosomes, for example, were discovered as early as 1978.

146. Ibid., 44.

147. Ibid.

148. Keller 2000, 145.

149. Davies and White 1996, 81–84. These family pedigrees would later be studied by Mark Skolnick and the team from Myriad Genetics as they mapped BRCA1.

150. Knudson 1971.

151. Everyone inherits two copies each of both BRCA1 and BRCA2.

152. Linkage analysis entails linking a known marker (for example, a polymorphism) on the genome and a particular phenotype (disease expression). If a group of individuals with a disease test positive for a known marker, researchers can assume that the gene for that disease resides close to that marker. Thus, researchers can narrow their search for the gene to a particular chromosome and to a particular location on that chromosome. If a family member has breast cancer but does not inherit a marker (due to the recombination of chromosomes during reproduction), then it is assumed that the gene is not close to that marker, and researchers rule out that particular region of the chromosome. In the 1980s, linkage analysis was used to map the Huntington gene in 1983 and the cystic fibrosis gene in 1985.

153. Jeff Hall et al. 1990.

154. They wrote: "Mapping genes for familial breast cancer is important because alterations at the same loci may also be responsible for sporadic disease" (ibid., 1684). In 1995 researchers observed damage to BRCA genes in sporadic tumors, and a debate ensued regarding how to interpret these findings. According to Lawrence C. Brody (personal communication, December 2000), it was believed that although BRCA plays some sort of role in sporadic cancer, it does so by a different mechanism than that which is involved in mutations that are inherited.

155. Miki et al. 1994. Autosomal dominant trait pattern simply means that one has only to inherit one bad copy of the gene to be at increased risk of cancer (as opposed to a recessive trait, in which two copies of a mutation must be inherited).

156. Wooster et al. 1995; Narod et al. 1991.

157. "Penetrance" refers to the proportion of individuals of a particular genotype that expresses its phenotypic effect in a given environment.

158. The guidelines expand considerably the number of women for whom a mutation can be predicted and who therefore should be offered screening. See ACOG 2009.

159. Sanz et al. 2010.

160. "Polymorphism" denotes a variation of a gene that may or may not be pathogenic. For clarity, I use the words "polymorphism" and "variation" to denote alterations of genes that may be benign and "mutation" to denote alterations that are not. For a study that exemplifies this particular research trajectory, see T. Smith et al. 2008.

161. Examples include variations of genes that affect how the body metabolizes carcinogens and genes that can repair the damage of such agents.

162. F. Collins et al. 2003, 7. This "vision" for genomics is certainly ambitious and arguably overstates just what genomics research is capable of accomplishing—the effect of pressure on scientists like Francis Collins (then director of

the National Human Genome Research Institute) to continually make the
case for public financing of this science. Nevertheless, the sequencing of genes
implicated in both rare and chronic diseases has meant that genomics has had
practical effects in the field of biomedicine even if falling short of being truly
revolutionary in scope. These practical effects have manifested themselves
primarily in the commercialization of genetic tests that reveal evidence of
inherited susceptibility to illness.

163. Parthasarathy 2007. Parthasarathy's book also provides an excellent comparative analysis of BRCA in the US and UK contexts.

164. The plaintiffs include the American Civil Liberties Union and the Breast
Cancer Fund.

165. For an in-depth thematic history of public discourse about genomics, see
Condit 1999b.

166. Lewontin 1991, 7.

167. Other genes thought to be responsible for inherited breast cancer at the time
included p53, STK11/LKB1, PTEN, MSH2/MLH1, ATM, WT1, and CYP17.

168. See, for example, Cowley 1993.

169. See Patterson 1987.

170. Cowley 1993, 46.

171. Quoted in Painter 1994.

172. Maugh 1994b; Saltus 1994.

173. The rough estimate currently is 1 in 300–500 persons.

174. Angier 1994. In the dozens of research articles I have read, I have yet to come
across a researcher making such a statement. Rather, researchers initially
claimed that although inherited breast cancer accounts for only a small percentage of the overall incidence, research in this area would be more broadly
significant because of its potential applicability to sporadic—that is, not inherited, adult-onset—breast cancer.

175. See Casamayou 2001 for a history of breast cancer activism in the 1990s.

176. See, for example, Chen et al. 1995 and M. Thompson et al. 1995.

177. See, for example, Sorelle 1995.

178. King's prepublication findings were first presented at the annual American
Society of Human Genetics conference in 1990 (Davies and White 1996, 1–4).

179. Steeg 1996, 223.

180. Davies and White 1996, vii.

181. S. Friend 1996, 16.

182. To some extent, the language of the geneticists themselves contributed to this.
The authors of the 1994 report that pinpointed the location of BRCA1 appear
to water down the many technological and methodological issues raised by
BRCA1's description: "The role of BRCA1 in cancer progression may now be
addressed with molecular precision. The large size and fragmented nature
of the coding sequence will make exhaustive searches for new mutations

challenging. Nevertheless, the percentage of total breast and ovarian cancer caused by mutant BRCA1 alleles will soon be estimated, and individual mutation frequencies and penetrances may be established. This in turn may permit accurate genetic screening for predisposition to a common, deadly disease. Although such research represents an advance in medical and biological knowledge, it also raises numerous ethical and practical issues, both scientific and social, that must be addressed by the medical community" (Miki et al. 1994, 71). Although some writers have claimed that the "newness" of genomics explains the lack of critical review by journalists (Rick Weiss and Susan Okie, personal communication, December 2000), the decline of investigative journalism has been affecting science reporting for some time. Many newspapers rely on reports from other papers or simply run wire-service stories. Even science writers at larger papers (and presumably with more resources) tend to rely on the testimony of the authors of research reports and other geneticists likely to corroborate the views of the study's authors. Moreover, press embargoes often limit the time available for writers to investigate stories, thus exacerbating the problem of relying on testimony from the authors of research reports or short lists of establishment experts that rarely change.

183. Painter 1994.

184. T. Friend 1994.

185. Maugh 1994b.

186. Rochelle 1994.

187. The HGP, for example, was often talked about in terms of tackling the massive amount of information in DNA. Metaphors regarding the search for BRCA1 were similar: "if the total genome was represented by the distance between Bangor, Maine, and Miami, Florida, the researchers had narrowed the location of the breast cancer gene to the greater New York City area" (Davies and White 1996, 100).

188. Cowley 1993.

189. Saltus 1994.

190. Ibid., referring to recommendations by experts at the Dana-Farber Cancer Institute.

191. The study initially garnering attention was Wolff et al. 1993. Krieger et al. (1994b) published results that failed to confirm a link between DDT, PCBs, and breast cancer. But Dewailly et al. (1994) confirmed the results of Wolff et al. I talk more about this research in chapter 5.

192. In the text of the article, the writer claims that the 1994 Krieger et al. study "seems to eliminate" DDT and PCBs as suspects in the historical rise of breast cancer (Maugh 1994a). Likewise, the *Atlanta Journal- Constitution* claimed the study found "no evidence" of a connection between the chemicals and breast cancer (Husted 1994).

193. Kolata 2002; "Breast cancer mythology on Long Island" 2002.

194. "Breast cancer mythology on Long Island" 2002. Kolata's science reporting has been so one-sided in favor of corporate interests that she herself has been investigated by other journalists. See Dowie 1998.

195. According to Tesh (2000) and Couto (1986), most groups get their information from scientific experts. Also, many activists go through various types of training, so as to not be completely dependent on professionals (P. Brown 1987, 83; National Breast Cancer Coalition Fund 2000). I found this to be true of the breast cancer activists I spoke with on Long Island, in particular Barbara Balaban. Balaban stays up to date by reading scientific reports regularly and has been a member of several science advisory panels.

196. Sutton 1996.

197. Geri Barish, president of One in Nine, is quoted in a 1995 article as being "thrilled" that scientists are studying the environment because "I've always felt the environment was the No. 1 cause of breast cancer, and probably other diseases as well" (Rabin 1995). A Suffolk County councilwoman is represented in a similar fashion: "It's very upsetting to us, because we really think something is going on" (Cimons 1999, A1). Kolata (2002), for example, opposes experts with specified credentials who dispute the theory that a certain class of chemicals known as organochlorines increases breast cancer risk with "advocates" of the theory whose perception of risk drives their activism.

198. Schemo 1994; Rabin 1995.

199. Tusiani 1998.

200. Ibid.

201. Sociologists and historians of environmental breast cancer activism have treated these "growth pains," including coalitions that are ever-shifting, as expected social phenomena. See Ley 2009; Klawiter 2008.

202. In what follows, I discuss the kind of bioethics research that is most likely to receive funding by the NIH and that is most likely to be represented in the mass media. For an excellent critique of bioethics and liberal humanism, see Murray 2006; Shildrick 2004.

203. Quoted in Koenig et al. 1998, 533 (ellipses in original).

204. According to my review of studies published through 2003, the topics funded were: genetic counseling and minority/diverse communities (six); genetic counseling in general (six); attitudes toward and interest in genetic tests (six); psychological effects of tests, including effects on family relations (seventeen); attitudes of physicians or physician education (three); employment or insurance discrimination (three); informed consent (five); testing of children (three); computer programs for counseling patients (nine); attitudes regarding preventive oophorectomy (two); telephone counseling (one); use of mammograms (one); behavioral risk factors (one); criticism of popular media coverage of breast cancer genetics (one).

205. Rothenberg 1997.

206. Ibid.

207. One must also consider the extent to which bioethicists have a professional interest in many of the solutions they offer. For example, they often identify the shortage of genetic counselors as a significant problem, and training more counselors would largely be the responsibility of bioethics departments. Thus, recognizing the ethical problems of genetic testing, while also arguing that tests should remain part of medical care, fulfills their responsibility to be watchdogs and guarantees the necessity of their services. For a critique of the profession of bioethics, see Elliott 2001.

CHAPTER 3

1. Queller 2008. Queller was also interviewed by National Public Radio and other outlets.

2. Tuttle et al. 2009.

3. See Happe 2006. Sunder Rajan 2006 makes a similar point.

4. American Cancer Society 2005.

5. National Cancer Institute 2011b.

6. American Cancer Society 2012.

7. Centers for Disease Control and Prevention 2006.

8. M. Goodman et al. 2003, 2676.

9. Gilda Radner, a victim of ovarian cancer, was a well-known American actor and comedian.

10. Piver et al. 1993a.

11. Piver et al. 1993b. The lifetime risk for women in the general population is 1.8 percent. See King et al. 2003, 643.

12. Gallion and Park 1995.

13. Piver et al. 1993a, 588.

14. Narod et al. 1991. See also Jeff Hall et al. 1990.

15. See, for example, H. Risch et al. 2001; Olopade and Artioli 2004.

16. Haber 2002; Kinney et al. 2005, 2509.

17. See, for example, Piver et al. 1993a.

18. Seltzer et al. 1995, 493.

19. Runowicz 1999.

20. Burke et al. 1997, 1000.

21. See, for example, Tobacman et al. 1982.

22. The full name of this cancer is primary peritoneal carcinoma or papillary serous carcinoma of the peritoneum (PSCP). The peritoneum is the tissue that lines the abdominal wall and covers most of the organs in the abdomen. For an overview of theories as to why PSCP is a risk for some women, including BRCA carriers, see Eisen and Weber 1998.

23. Kauff et al. 2002, 1613.

24. Rebbeck et. al. 2002, 1621.

25. Levine et al. 2003, 4226.
26. Rosen et al. 2004, 285.
27. Garber and Hartman 2004, 978.
28. "Ovariotomy" is the term used early in this surgery's history, but it is not used now. "Ovariectomy" was and is still used, and historians have noted that it should be distinguished from "oophorectomy," which denotes ovary removal for prophylactic reasons. According to this distinction, I will use "ovariectomy" to denote removal for treatment of ovarian disease (including, but not limited to, ovarian cancer) and "oophorectomy" to denote removal for preventive purposes.
29. Moscucci 1990, 6.
30. Smith-Rosenberg and Rosenberg 1999.
31. Sengoopta 2000.
32. Rutkow 1999, 902. See also Utian 1999; McMahon 2000. The first recorded attempt is dated 1771 (McMahon 2000, 401).
33. Barker-Benfield 1976, 121. One important limitation of the history that I have narrated here is that it does not detail which bodies were more available than others, a limitation of the historical evidence available. There is some anecdotal evidence that slave women were used as experimental subjects; according to one historical account, the first successful abdominal hysterectomy with bilateral salpingo-oophorectomy was performed in Knoxville, Tennessee, in 1856 on a "negress" servant (Jason Hall and Hall 2006). Many of the surgeons mentioned in the histories I consulted practiced in the South, which would suggest that black women were the unfortunate victims of experimental surgery. There does seem to be evidence that oophorectomies on poor women and women of color played a role in the development of reproductive technology, such as in vitro fertilization (Corea 1985; see also Marsh and Ronner 2008).
34. Moscucci 1990, 144. In the late 1880s, critics would compare ovariectomy to vivisection, suggesting that women were being used as experimental subjects in torturous procedures, an image that further suggested comparisons with Jack the Ripper.
35. Ibid., 140.
36. The mid-1800s were an important time for the field. In 1852 Marion Sims opened the first hospital for a specific group—women—in New York. His achievements are generally credited with "raising gynecology to a respected medical specialty" (Barker-Benfield 1976, 89).
37. Thiery 1998; Moscucci 1990; Dally 1991.
38. Sengoopta 2000, 451–52.
39. Moscucci 1990, 134.
40. Barker-Benfield 1976, 82; Dally 1991.
41. Studd 2006b, 414.

42. Ibid., 412.
43. Moscucci 1990, 156. See also Thiery 1998, 243; Dally 1991; Sengoopta 2000. "Cyrrhosis" is an alternative spelling for "cirrhosis." Presumably, in the case of ovaries as in the case of the liver, cyrrhosis was the term used to denote scarring of the tissue.
44. In the United States, it became known as "Battey's operation" (after Robert Battey of the state of Georgia); in England, it was "Tait's operation" (after Lawson Tait of England) (Studd 2006b, 412; Moscucci 1990), although Alfred Hegar of Germany is also credited for performing the surgery in 1872. In 1872 Robert Battey "presented before the American Gynecological Society four indications for surgery: (1) when the absence of a uterus threatened a patient's life; (2) when it was impossible to restore an obliterated uterine cavity or vaginal canal by surgical means; (3) in cases of uterine or ovarian insanity or epilepsy; (4) in cases in which monthly periods produced prolonged mental and physical suffering" (Morantz-Sanchez 1999, 95). The openness with which Battey defined appropriate conditions for surgery suggests that he was performing surgeries often "for a range of vague mental and physical symptoms" (ibid).
45. Literally, ovary-induced mania.
46. Excessively painful menstruation.
47. Studd 2006b, 413.
48. Ibid.
49. Barker-Benfield 1976, 121. William Goodell, a professor at the University of Pennsylvania, advocated performing an oophorectomy on all insane women because he believed insanity was a heritable trait (Dally 1991).
50. Quoted in Sengoopta 2000, 426. Chapter 2 discusses the topic of eugenics in greater depth.
51. See Ehrenreich and English 1978. For general histories of women and feminism, see Freedman 2002; Evans 1989.
52. Smith-Rosenberg and Rosenberg 1999, 115.
53. Quoted in Barker-Benfield 1976, 88.
54. Smith-Rosenberg and Rosenberg 1999, 111.
55. Ehrenreich and English 1978, 145.
56. Smith-Rosenberg and Rosenberg 1999, 112.
57. Morantz-Sanchez 1999; Sengoopta 2000.
58. Smith-Rosenberg and Rosenberg 1999, 113.
59. Moscucci writes: "In the last quarter of the nineteenth century, as Britain embarked upon its most ambitious programme of colonial expansion, imperialism stood little chance of success without motherhood" (1990, 159).
60. Resistance to ovariectomy is thus an early instance of the strategic appropriation of popularly held ideas about femininity and womanhood in order to bring about reform (in this case, medical reform). Historians have described

the cult of motherhood and reform campaigns. See Ehrenreich and English 1978; Freeman 1998; Evans 1989.

61. Thiery 1998, 245. The recognition that the hormonal function of ovaries could be not replicated using methods available at the time also played a role in the demise of the procedure (Moscucci and Clarke 2007).

62. Thiery 1998, 245. See also Moscucci and Clarke 2007.

63. "Involution" means the decrease—in size and functional activity—of an organ whose role in the body is temporary or confined to certain periods of life.

64. Quoted in Moscucci and Clarke 2007, 183.

65. "Benign" means "not cancerous"—hence, the possibility of "benign pathology." As recently as 2000, hysterectomy was the second most frequently performed major surgery in the United States. The majority of these operations, critics have argued, are unnecessary. In a survey of case studies in which doctors recommended hysterectomy in order to alleviate disease symptoms (such as those associated with endometriosis), researchers concluded that the vast majority of recommendations were based on flawed or simply nonexistent knowledge regarding the patient's condition and possible alternative treatments (Broder, Kanouse, Mittman, and Bernstein 2000, 199).

66. According to Parker et al. (2005b), the most frequent reason for prophylactic oophorectomy with hysterectomy for benign disease is to decrease risk for ovarian cancer. Each year 1.2 million oophorectomies are performed. Half are bilateral, and half of those are performed for prophylactic reasons (Mayo Clinic 2006). See also ACOG Committee on Practice Bulletins 1999.

67. It is unclear why these rates are so different, given the fact that attitudes about oophorectomy among physicians seem to be the same. Parthasarathy, while not addressing the topic of oophorectomy directly, does offer some clues in her comparative analysis of BRCA in the United States and the United Kingdom. She says that women in the United States are more likely to see themselves as diseased simply by being a mutation carrier, largely the result of marketing discourse by Myriad Genetics. In contrast, in the United Kingdom, testing is regulated differently as part of the country's National Health Service. Women are counseled when a significant family history is present, whether or not this involves BRCA mutations. These "broad public health concerns," says Parthasarathy (2007, 203), also account for why Tamoxifen, given its serious side effects, was never routinely offered to British women as a preventative.

68. Parker et al. 2005b.

69. Ibid., 219

70. Moscucci and Clarke 2007, 182.

71. Olive 2005, 214. Another example of recommending oophorectomy during abdominal or pelvic surgery is Seltzer et al. 1995, 496.

72. Eisen, Rebbeck, Wood, and Weber 2000, 1982. Huisman recounted the struggle over medical dogma in the Netherlands: "About two decades ago we

had the same discussion in the Netherlands. At that time it was customary to remove the ovaries whenever a woman aged ≥45 years was operated on by a gynaecologist. We had a hard time convincing our teachers that ovaries at the time of the menopause, and for a length of time thereafter, were still a prized possession. The reasoning of our elders was that once the ovaries were removed, no grim disease such as ovarian cancer (a cancer that is almost always beyond control at the time of detection) could develop" (1997, 204).

73. Runowicz 1999.
74. See LeWine 2005; Olive 2005.
75. Piver and Wong 1997, 206.
76. Davy and Oehler 2006, 167.
77. Ibid. The authors put it quite forcefully: "In the absence of a known BRCA1 or BRCA2 mutation, or a very strong family history of ovarian and/or breast cancer, even the most fanatic gynecological oncologist would not suggest to a woman with low risk factors that she have bilateral oophorectomy to decrease the possibility of ovarian cancer" (167). Of course, what is considered "fanatical" changes over time, as the history of prophylactic oohorectomy suggests.
78. Piver and Wong 1997, 206. See also Studd 2006a.
79. According to a 1999 ACOG practice bulletin, reoperation rates for women who undergo hysterectomy can be as high as 5 percent, with pain in the retained ovary being the most common reason (ACOG Committee on Practice Bulletins 1999, 195). Also, women who have a hysterectomy tend to be younger, and risks associated with surgery increase with age. Davy and Oehler qualify this: "Repeat operations on usually much older women have their technical difficulties and complications. Furthermore, the procedure creates significant stress and anxiety for the patient until the ovarian malignancy is excluded. Thus, many operations on postmenopausal ovaries could be avoided by oophorectomy at the time of hysterectomy" (2006, 168). Studd cites research on women with endometriosis, for whom reoperation rates can be as high as 40 percent (2006a, 165).
80. Piver and Wong 1997, 205–6.
81. Parker et al. 2005b.
82. Armstrong et al. 2004. The 300,000 figure cited earlier does not distinguish between women with or without BRCA mutations.
83. Parker et al. 2005b, 219.
84. Ibid., 222.
85. Ibid., 223. Menopause increases the risk for heart disease—hence the conclusion that women should retain their ovaries as long as possible. Nevertheless, much more research is needed to fully understand the function of postmenopausal ovaries. The study by Parker's team seems to suggest that the risk for heart disease is even higher when menopause is surgically induced at a young age. Complicating this further are data released from the Women's Health

Initiative after the Parker et al. study that suggests HRT can *increase* the risk of cardiovascular disease in postmenopausal women. See Toh et al., 2010.

86. For example, oophorectomy resulted in a substantial increase in coronary heart disease for postmenopausal women, a risk not seen in women undergoing natural menopause (Colditz et al. 1987).

87. S. Shapiro 2006, 163.

88. Davy and Oehler 2006, 168.

89. Ibid.

90. Studd 2006a, 166; Studd 1989.

91. Studd 2006a, 166.

92. Parker et al. (2005b) duly note this.

93. LeWine 2005.

94. Olive 2005, 215

95. Parker et al. 2005a, 1107. Parker and coauthors were responding to two letters to the editor that criticized the research, using arguments of Studd and others: that HRT did in fact have health-promoting properties despite the results of the Women's Health Initiative and the Nurse's Health Study; that the risk for future operations is high enough to warrant earlier surgery; and that statins and biophosphonates can reduce risk from heart disease and osteoporosis, respectively (Dandolu and Hernandez 2005; Wallach 2005).

96. Parker et al. 2006, 396.

97. For evidence that breast cancer rates dropped after publication of the Women's Health Initiative results, presumably because fewer women were taking hormone replacement therapy, see Chlebowski et al. 2009.

98. Parker et al. 2006, 397.

99. Rocca et al. 2006.

100. Some of these cancers may have been the result of receiving estrogen-only therapy (ibid.). One physician speculated that "the increased risk of death from breast and uterine cancers may have been due to how estrogen receptors function in the body's different organ systems . . . We don't understand completely the function of estrogen receptors at this point. It's *intriguing* that keeping the ovaries might reduce the risk of getting cancer" (quoted in Lindsey 2006, 24; emphasis added). Further highlighting how little is known about estrogen function after menopause, the physician also theorized that ovaries play an important role in controlling hot flashes and preventing osteoporosis. See E. Martin 1987 for a feminist analysis of the treatment of menopause.

101. Rocca et al. 2006, 826.

102. Ibid., 827.

103. The authors do not draw a firm conclusion as to whether HRT can mitigate the harmful effects of prophylactic oophorectomy before the age of forty-five. They do caution that oophorectomy could be a marker of as-yet-unidentified disease and/or genetic susceptibility that could predispose women to early

death. In other words, the women more likely to undergo oophorectomy are more likely to be predisposed to the several diseases observed in the study; therefore, oophorectomy is not necessarily the causal factor. The problem with this logic is that history shows that oophorectomy is overprescribed for reasons that have nothing to do with factors that may predispose women to get the surgery.

104. See Lindsey 2006 for interviews with several physicians.

105. ACOG Committee on Practice Bulletin 2008, 237.

106. Eisenberg 2006.

107. E. Martin 1987.

108. Hickey, Ambekar, and Hammond 2009, 139.

109. In its 1999 practice bulletin on prophylactic oophorectomy, ACOG rates the following recommendation "Level C," meaning "based primarily on consensus and expert opinion": "Prophylactic oophorectomy should be considered for select women at high risk of inherited ovarian cancer" (ACOG Committee on Practice Bulletins 1999, 197). By 2008 the recommnedation had changed to "Level B," meaning "limited or inconsistent scientific evidence": "Bilateral salpingo-oophorectomy should be offered to women with BRCA1 and BRCA2 mutations after completion of childbearing" (ACOG Committee on Practice Bulletins 2008, 237).

110. If the decision model is valid, wrote one critic, "the public health consequences would be major" (S. Shapiro 2006, 163).

111. Haber 2002, 1660.

112. Ibid.

113. S. Miller et al. 2005, 655.

114. A. Goodman and Houck 2001, 93.

115. Liede et al. 2002, 1570.

116. Colgan et al. 2001, n.p. Although the researchers relied on self-reports of family history—a potentially significant limitation of the study—the Cass et al. study (2003) and others like it nevertheless point to a tendency among cancer geneticists to delink BRCA-related cancers from cases associated with cancer syndromes. Other studies have shown just what some extrafamilial variables might be. For example, in 2001 a large study of women with ovarian cancer revealed that women with nonmucinous tumors (those that are poorly differentiated and much harder to treat) were significantly more likely to test positive for BRCA mutations; the researchers concluded that it is "reasonable to offer genetic testing to all women with invasive nonmucinous ovarian cancer" (H. Risch et al. 2001, 708). Ethnicity has also emerged as a marker for genetic risk. In one study of Jewish women with ovarian cancer, the researchers concluded that "in light of the high probability of BRCA mutations in Jewish patients with ovarian carcinoma, we currently discuss genetic testing with all newly diagnosed, Jewish patients with ovarian/peritoneal or fallopian

tube carcinoma" (Cass et al. 2003, 2193–94.). Histology and ancestry have thus become evidence of inherited risk, evidence that does not necessarily appear in conjunction with family history of the disease. Even age of onset is no longer a reliable predictor of a mutation, as women with BRCA2 mutations do not develop ovarian cancer at an appreciably younger age than women in the general population.

Many variables, then, independent of family history or age of onset, are grounds for conducting genetic screening. For women with ovarian cancer, genetic tests are potentially valuable because the information they reveal may affect treatment options: women with these mutations may respond differently to medication, and some are even expected to have better or worse chances of surviving the cancer (Rubin 2003). How their risk is retroactively defined, however, may have implications for family members. When a woman tests positive for a mutation, family members—even those individuals whose family would not meet the definition of a cancer syndrome family—may feel pressure to be tested. Although test results are typically not disclosed without the consent of the woman tested, many clinicians feel that women have a responsibility to tell family members who may need to avail themselves of risk-prevention strategies such as prophylactic surgery (see, for example, Cass et al. 2003). Thus, although the number of women suspected to test positive for a BRCA mutation remains comparatively low (5–10 percent of ovarian cancer cancers are thought to be hereditary), of those women, the proportion who will be tested may increase. Once tested and marked as at risk, these women become targets of interventions like prophylactic surgery.

117. For risk scholarship that demonstrates this, see Beck 1992. See also chapter 5.
118. The analysis is sometimes carried out using what is called a protein-truncation test, when clinical data are unavailable for a mutation but researchers need to know whether it is "deleterious" (see, for example, H. Risch et al. 2001). More sophisticated techniques are being developed to determine this, leading to the reclassification of mutations in the Breast Information Core database (see Sanz et al. 2010).
119. H. Risch et al. 2001.
120. King , Marks, and Mandell. 2003.
121. I will defer for now the very important question as to whether the privileged status of the BRCA mutation is possible because actual bodily matter is collected for analysis.
122. See, for example, Treichler, Cartwright, and Penley. 1998.
123. Levin 1993.
124. Spear et al. 1999, 550.
125. Eisen and Weber 1998, 798.
126. Bourne et al. 1991. There is no question that if genetics research can develop more effective, less violent cancer treatments, perhaps by detecting early

molecular changes preceding tumor development and reversing their course, then genetics will contribute in very positive, significant ways to women's health care. The point of this chapter is merely to describe what scientific practices have emerged in the meantime.

127. Several studies have suggested that because cancer is often detected at the time of prophylactic surgery, the procedure in effect functions as a screening method, thus overcoming the limits of existing screening technologies.

128. Kauff and Barakat 2004, 277.

129. Davy and Oehler 2006,167.

130. Van Nagell 1991, 91.

131. The Prostate, Lung, Colorectal and Ovarian (PLCO) Cancer Screening Trial collected data from 2002 to 2008 for about 30,000 women undergoing screening such as CA-125 and transvaginal ultrasound. Early results published found that screening does detect early-stage ovarian cancer but also results in many false positives (Buys et al. 2005).

132. Of course, I am not suggesting conscious strategizing on the part of geneticists to use heredity for the same ends as gynecologists did in the 1800s, only that the ways in which women's bodies are available for intervention have changed.

133. Even if one remains suspicious of the Parker model, the Rocca study nevertheless calls into question the practice of prophylactic surgery for premenopausal women—a finding that was not, apparently, as controversial. It can also be argued that the kind of defensiveness provoked by the Parker study is almost always a response to any challenge to medical dogma. Kuhn's (1970) thesis on scientific change is relevant here.

134. Eisen and Weber 1998.

135. Liede et al. 2002, 1575.

136. Karlan 2004, 520. One researcher concluded that HRT merely negates the reduction in breast cancer risk that oophorectomy provides, but this claim is unlikely to be persuasive to women with breast-ovarian heredity syndrome.

137. Estrogen-only HRT has not been shown to increase risk for breast cancer. It is unclear why.

138. Kauff and Barakat 2004, 277–78.

139. Kauff et al. 2002.

140. Grann et al. 2002.

141. King et al. 2001.

142. Grann et al. 2002.

143. Cass et al. 2003.

144. Schrag, Kuntz, Garner, and Weeks 1997.

145. Rebbeck et al. 2002, 1621.

146. Ibid.

147. In 2006, 315,930 women died from heart disease; this represents one out

of every four women with the condition (Centers for Disease Control and Prevention 2012b). In contrast, the National Cancer Institute (2011a) predicted that five years later, 39,500 women would die from breast cancer, out of 230,500 diagnosed with the disease.

148. Fry, Rush, Busby-Earle, and Cull 2001; Hurley et al. 2001.

149. Quoted in L. Newman 2001.

150. Domchek et al. 2006, 223.

151. H. Risch et al. 2001.

152. See, for example, Haber 2002.

153. Haber 2002, 1661.

154. Hallowell et al. 2001, 683.

155. Kauff and Barakat 2007, 2926.

156. Ibid., 2922.

157. Ibid., 2922.

158. Huggins and Dao 1953, 1389. For an overview of why oophorectomy is once again being considered for treatment of breast cancer, see Love and Philips 2002.

159. Batt 1994. These were women who were undergoing oophorectomy as a form of breast cancer treatment.

160. Batt 1994.

161. This is changing, however. Recent data show a significant increase in mastectomy for women diagnosed with breast cancer. See Tuttle et al. 2009.

162. See, for example, L. Hartmann et al. 1999.

163. Contrary to this, many articles on prophylactic mastectomy for BRCA mutation carriers have noted that the development of more sophisticated plastic surgery techniques may make this surgery more attractive: "Despite the paradox of treating at-risk patients with mastectomies while treating breast cancer patients with breast conservation, prophylactic mastectomy in BRCA-1 and BRCA-2 patients may prove to be a rational choice. If the patient develops invasive breast cancer, she may require chemotherapy and be at risk for the development of life-threatening metastases even if she is adequately treated locally with breast conservation. For these patients, bilateral prophylactic mastectomies may reduce the risk of distant disease. Reconstruction with bilateral TRAM flaps at the time of oophorectomy and hysterectomy is technically possible and can yield excellent results" (Spear et al. 1999, 552).

164. Eisen, Rebbeck, Wood, and Weber 2000, 1981. See also Rebbeck et al. 2002.

165. This was not always the case—recall that in the 1800s resistance to oophorectomy relied on the argument that the surgery desexed women, robbing them of their gendered identity; to some extent, this argument is also in circulation today. See, for example, Elson 2004. For a cultural history of the breast, see Yalom 1998.

166. For an excellent critique of the estrogen hypothesis as it pertains to breast

cancer risk, see Krieger 1989. I discuss Krieger's essay in greater detail later, when I analyze the relationship between race, reproduction, and cancer risk. In a related vein, see Fausto-Sterling's 1985 work on how flawed assumptions about estrogen have negatively influenced biological research about women.

167. Rebbeck et al. 2002, 1620.

168. Conversely, a BRCA1 carrier may want to undergo oophorectomy sooner rather than later, as her diagnosis will probably occur earlier than diagnoses for women in the general population.

169. Cass et al. 2003.

170. Ibid., 2193. Further evidence of this comes from Olopade and Artioli: "Of interest is that there have been no events in the 124 women who had oophorectomy by age 35 years, which suggests that the timing of oophorectomy may be important" (2004, S7).

171. Kauff and Barakat 2004, 277. For research regarding the protective effect of multiple, early pregnancies for women at risk for BRCA-related breast cancer, see Marquis et al. 1995; Holt et al. 1996; Jensen et al. 1996.

172. Heffner 2004.

173. Aliyu et al. 2005.

174. Faludi (1991). Consider the cyclical pathologization of the single and/or working mother in US public discourse, most often featured in front-page exposés in magazines like *Time* and *Newsweek*. The association of mental illness with single life, or with what is known as "having it all" (that is, both working and having a family) is a feature of biopolitics that Faludi famously identified as a key part of the backlash against US women in the wake of the feminist movement.

175. In an effort to make up for the dearth of knowledge regarding health in the lesbian community (many epidemiological studies do not note sexual orientation, so, for example, no one knows whether incidence rates of breast cancer are higher or lower for lesbians than for heterosexual women), Suzanne Haynes of the National Cancer Institute reported that lesbians were at a much greater risk of breast cancer than heterosexual women. Haynes did not report on incidence of breast cancer, however. Her methodology entailed generalizing lesbian lifestyles and noting the degree to which risky behaviors were more typical among lesbians. Based on assumptions such as fewer pregnancies and greater alcohol consumption, Haynes announced that as many as one in three lesbians is at high risk. The report garnered significant news coverage, which at times used the imagery of "plague" ("Cancer danger for lesbians" 1993; Raeburn 1993; Selvin 1993). Many of the news reports failed to mention that more research was needed to study these presumed risk factors as well as determine actual incidence of breast cancer. A study led by Stephanie Roberts of Lyon-Martin Women's Health Services in San Francisco found significant differences in some risk factors (S. Roberts et al. 1998). In terms of incidence, however, Roberts

and coauthors found that there was no actual difference between lesbians and heterosexual women. Yet headlines about the study read "Breast Cancer Risk Higher in Lesbians" (White 1998) and "Breast Cancer Risk Factors Appear Higher in Lesbians" (Kim 1998). The 1998 study also prompted more general claims about pregnancy and breast cancer risk, such as: "Women who have never been pregnant are about six times more likely to develop breast cancer" ("Breast Cancer Risk Likely Higher for Lesbians" 1998). Many of these news reports were reprinted in multiple newspapers.

176. See *International Union v. Johnson Controls, Inc.*, 499 U.S. 197 (1991). By one estimation, as many as twenty million workers were at risk because of so-called fetal protection policies by employers in the mid-1990s (Samuels 1996, 209). Although the *Johnson Controls* ruling was favorable to women, the actual outcome was not. Employers have used the case to deny special accommodations for women employees who become pregnant (for example, maternity leave and protection from exposure to chemicals), arguing that special treatment of female workers would violate the spirit of *Johnson* (Samuels 1996).

177. Butler 1990.

178. Ibid., 7.

179. In an interview, Butler put it this way: "Now it seems to me that, although women's bodies generally speaking are understood as capable of impregnation, the fact of the matter is that there are female infants and children who cannot be impregnated, there are older women who cannot be impregnated, there are women of all ages who cannot be impregnated, and even if they could ideally, that is not necessarily the salient feature of their bodies or even of their being women. What the question does is try to make the problematic of reproduction central to the sexing of the body. But I am not sure that is, or ought to be, what is absolutely salient or primary in the sexing of the body. If it is, I think it's the imposition of a norm, not a neutral description of biological constraints" (quoted in Osborn and Segel 1994, 33).

180. The change in Fausto-Sterling's thinking is interesting in this regard. After initially suggesting that the number of sexes be expanded from two to five, she has since called for replacing the male-female sex binary with a notion of sex as multiple points on a continuum. Intersex advocates have called for medical reform so no surgeries would be performed prior to the age of consent, save those deemed medically necessary (a potentially problematic exception, but one that is necessary in order to account for the physiological problems that sometimes accompany the biological development of the intersex). There would be, in a post-intersex world, an unlimited number of sex categories and presumably an unlimited number of genders. Although idealistic and not entirely mindful of Butler's important argument about the enforcement of gender norms (even when sex appears uncertain), by disarticulating biological sex from the categories "male" and "female,"

Fausto-Sterling arguably disarticulates the categories from mere reproduction with her notion of a continuum (2000a, 2000b).

181. A. Davis 1981; Solinger 2001; D. Roberts 1997a; P. Collins 2000.

182. D. Roberts 1997b; Solinger 2001.

183. Another result is differential practices of prenatal genetic counseling. Genetic counselors, for example, have admitted to being more likely to recommend abortion for carriers of the trait for sickle cell anemia, reasoning that the perceived inability by women who carry the gene to care for the child renders medical care futile (Rapp 1998).

184. D. Roberts 1997b; Reilly 1987.

185. D. Roberts 1997b. Norplant is no longer on the market, but it is an example of how insidiously sterilization can be achieved.

186. Mosher and Jones 2010. There are many reasons for this, including cultural influences (for example, women tend to adopt the types of birth control practices they are most familiar with), racist practices of physicians and family planning clinics that offer sterilization for blacks at rates far higher than for whites, financing that makes sterilization more affordable than other types of birth control, and the understandable desire of black women to have as little contact with the medical community as possible.

187. M. Goodman et al. 2003, 2684.

188. Ibid., 2618.

189. Parham et al. 1997, 816–17.

190. Randall and Armstrong 2003, 4203.

191. Palmer et al. 2003.

192. Krieger 1989.

193. Martin et al. 2012. Recall, too, that lower ovarian cancer rates are thought to be explained in part by higher rates of sterilization among African American women.

194. Olopade 2004; Armstrong et al. 2005; Lipkus, Iden, Terrenoire, and Feaganes 1999; Kinney et al. 2001; Thompson et al. 2002.

195. It is beyond the scope of this chapter to speculate about other reasons why blacks refuse testing, such as concerns over the trade-off between genomics and social and environmental justice. Researchers investigating attitudes toward testing and associated levels of uptake do not seem to allow for reasons other than culture and the lack of uniform, national safeguards to protect privacy and employment (research on cultural differences in attitudes does have merit, although it is always carried out by comparing blacks and whites, which results in gross generalizations about both groups). I am sympathetic with Dorothy Roberts's (1997b) argument that blacks' skepticism of genomics is just as likely to be rooted in concerns over social justice as it is in fear of the medical community or of medicine more generally. (Of course, given the persistence of racism in medicine, it is not at all unjustified for blacks to be

fearful.) Another limit of research on attitudes about BRCA is that researchers reduce all fear to the legacy of the Tuskegee syphilis study without any
acknowledgement of persistent racism, which is seen most clearly in the
statistics on health disparities and environmental racism—in spite of the fact
that women report environmental factors such as pollutants from chemical
plants as perceived causes of breast cancer (Kinney et al. 2005). Roberts documents that when it comes to other medical technology, blacks wholeheartedly
embrace radical intervention, as in the case of life-support measures. To
"ensure the just use of knowledge about genetics," she suggests, requires that
researchers begin with the experience of black women's lives to understand
how racism manifests itself as particular biological effects and how concerns
for social justice shape responses to the integration of genomics into routine
medical care. Resistance should be defined not as only the legacy of historical abuses of black patients by the medical community, but as an important
critique that can and should redirect the conduct of basic science and the
application of its knowledge claims. "Our task," Roberts argues, "is not simply
to increase knowledge about genetics, but, more importantly, to ensure the
just use of knowledge about genetics" (1997b, 972).

196. Huo and Olopade 2007.
197. Thompson et al. 2002.
198. Olopade 2004.
199. Ibid., 1685.

CHAPTER 4

1. Grady 2006b. The authors of the study—hailing from the fields of epidemiology, oncology, and genetics—concluded that when specific tumor types are
   taken into account, African American women, especially if they are premeno-
   pausual, are more likely to be diagnosed with deadlier forms of the disease
   than older African American women or non–African American women,
   regardless of age (Carey et al. 2006).
2. Grady (2006a) was responding to criticisms she received via e-mail after her
   story was published. For examples of scholarship on the problem of race in
   genomics research, see Reardon 2004; Stevens 2002, 2003 2008; Fausto-Sterling 2004; Fujimura and Rajagopalan 2011; Duster 1990.
3. M. Brown, Carnoy, Currie, and Duster 2003.
4. Fields 1990, 114. In the nineteenth century, race explained what Fields
   describes metaphorically as "terrain," a social landscape in which most people
   could take for granted the notion of liberty. Slavery needed to be explained
   away as an anomaly to fit with the new terrain; race served this function. Ideology, then, is a vocabulary that fits the social landscape in which one lives. It
   is a vocabulary that (descriptively) explains to the masses (white people) why
   the social landscape is configured in a particular way. That people today still

believe in "race" is evidence that "people are more readily perceived as inferior by nature when they are already seen as oppressed" (106).

5. Clinton, Blair, Collins, and Venter 2000.

6. Several of the early BRCA family studies detected the 185delAG mutation. See, for example, Struewing et al. 1995b. The letters and numbers "185delAG" signify the exact alterations of base pairs, which in turn explain the improper functioning of the allele.

7. Struewing et al. 1995a This raised a whole host of issues regarding informed consent; see Goldgar and Reilly 1995.

8. Ibid.

9. Ibid., 199.

10. Struewing et al. 1997. These variations in risk are rarely mentioned in the popular media. See chapter 2.

11. Keoun 1997, 8. Among the countries mentioned in this article are Norway (one mutation is common in the eastern part of the country and another in the west), Iceland, Russia, the Netherlands, and Finland; also mentioned are persons identified as Japanese, Swedish, Italian (specifically, from Tuscany), African American, and Scottish.

12. "Caucasian" is not normally considered an "ethnic" category. The articles I studied for this chapter use the terms "race" and "ethnicity" interchangeably or simply say "race/ethnic." It is typical for these authors to say at the beginning of the article that they are studying "race/ethnicity" and later name specific groups. But they do not then say which of those groups are ethnic groups and which are racial groups—it is up to the reader to infer the difference. I do not bring this up to suggest that shared linguistic conventions would improve matters; indeed, the fact that racialization occurs despite the impressive array of word choices appears to support my point that race is a field-invariant habit of mind or ideology.

13. Gao et al. 1997, 1233.

14. Ibid., 1234.

15. Ibid., 1235.

16. Ibid., 1236.

17. Ibid. See also Gao et al. 2000a; Pal, Permuth-Wey, Holtje, and Sutphen 2004. Earlier studies were inconclusive as to whether or not BRCA mutations would be found to a significant degree among African American women tested. A study in 1998, for example, found no mutations among any of the black subjects (B. Newman et al. 1998).

18. Gao et al. 1997, 1235.

19. Ibid., 1233. Other examples of using rates to justify research include M. Hall and Olopade (2006, 2201). Medullary carcinoma is a particular kind of tumor. It typically begins in the milk ducts and tends to be clearly differentiated from surrounding healthy tissue.

20. Gao et al. 1997, 1233.

21. For a review of different meanings of the term "health disparities" and their policy implications, see Carter-Pokras and Baquet 2002. Geneticists have also expanded the definition of "health disparities," complicating the picture even further. For example, Olopade (2004) considers lack of data regarding risk and BRCA mutations in black women, as well as lack of equal access to genetic testing, counseling, and prevention, to be a case of health disparities in medicine.

22. "Class" is also the source of health disparities (Levins and Lewontin 1985), but it is not typically employed in US epidemiological research. For a corrective, see Nancy Krieger's scholarship (for example, 2001).

23. To take just one example, Trock (1996) explicitly calls for BRCA research on African Americans because of what founder mutation research revealed for people of Ashkenazi descent.

24. Newill 1961. Religion at that time was listed as a cultural variable that could affect breast cancer risk.

25. Religion was sometimes linked with lifestyle and reproductive patterns. See Newill 1961.

26. See, for example, Slovut 1995. This was not the case: the study did not establish what risk might be associated with the mutations, only the frequency with which they were present in that particular population. Only later studies would establish risk.

27. Egan et al. 1996, 1645.

28. Struewing et al. 1997.

29. Trock 1996. It is preposterous to claim that any one study or group of studies can completely control for confounding variables; I develop this critique later in the chapter.

30. Ibid., 196.

31. Amend, Hicks, and Ambrosone 2006, 8327.

32. Epstein 2007. Epstein provides a thorough history of the emergence of the inclusion paradigm, or, the requirement that federally funded research include study subjects from different races and sexes.

33. Shim 2002, 132.

34. Ibid. This is also a problem for public health, as I discuss in chapter 5. When these studies confirm the existence of health disparities (for example, young black women are at greater risk for breast cancer than young white women), researchers assume that the role of race has been elucidated as well as confirmed. But how race exerts its effects has not actually been explained at all.

35. Kaufman, Cooper, and McGee 1997, 627.

36. I talk more about this research below in the chapter.

37. In February 2011 I was part of a panel on race and science at my home institution, along with a geneticist who had participated in the Chicago-based studies of social isolation and breast cancer risk among black women. When

I brought up these studies, she corrected my interpretation that the studies imply that we should consider interventions at the policy level to ameliorate health disparities. According to her interpretation, it was not social isolation per se that had led to higher glucocortisoid levels, but the *perception of* social isolation, a highly individualized assessment of both cause and level of intervention.

38. Shim writes: "Related to such [individualizing] practices is the assumption that characteristics of one's social environment, as incorporated in the multifactorial model, are exogenous to the individual—that one's circumstances are taken as a given, as if individuals were dropped into a set of conditions that are not socially constructed nor patterned. Modern epidemiology, despite growing efforts to develop more ecological models that acknowledge causation at multiple levels, remains, therefore, predominantly concerned with the identification of individual-level risk factors, earning it the label of 'risk factor epidemiology'" (2002, 133).

39. Chowkwanyun 2011, 254. Chowkwanyun offers an account of race similar to what Stevens calls a "synchronic" view. See Stevens 2008.

40. Krieger 1999.

41. Krieger 2001, 668. Punctuation errors in original.

42. I discuss the work of these critics below in the chapter.

43. Toniolo and Kato 1996. The Ashkenazi studies garnered considerable media attention, and various advocacy organizations voiced concern that the research implied Jewish women were genetically defective and, as a result, were at greater risk for breast cancer than other women. The Jewish Women's Coalition on Breast Cancer was formed as a result. The researchers themselves made it a point to say that Jewish women are at no greater risk for breast cancer than women in the general population. In an interview with the *Journal of the National Cancer Institute*, for example, Lawrence Brody claimed there are no data to show that rates of breast cancer are higher among Jewish women (Nelson 1998).

44. For an example of "Jewish" breast cancer gene talk, see "Ashkenazi Jewish women" 1995. Science writers and researchers alike invoked the Ashkenazi BRCA research to explain higher rates of breast cancer on Long Island, where epidemiologists have suspected the influence of environmental pollution (see, for example, Hotz 1995). When the Long Island Breast Cancer Study (which I will discuss in more detail in chapter 5) was in the planning stages, geneticists approached the lead researcher to see if the study could include testing the subjects for BRCA mutations (Marilee Gammon, personal communication, September 2000). This did not happen, largely because of logistical problems associated with genetics research such as genetic counseling (ibid.) (It is worth noting, too, that in the Long Island Breast Cancer Study report, the authors note that known risk factors, including genetic predisposition, do not explain

the elevated rates in New York—hence, the necessity of the kind of epidemiological research that the study required (eee National Cancer Institute
and National Institute of Environmental Health Sciences 2000, 3).Also, when
researchers have examined the association between age and risk among
mutation carriers, they found that risk for individual women increases over
time: "Nongenetic factors may significantly influence the penetrance even of
high-penetrance mutations. Breast cancer risk in women born before 1940 is
high, but risk is even higher for women born after 1940" (King, Marks, and
Mandell, 2003, 645).

45. Mullineaux et al. 2003.

46. Siegel 1996.

47. John et al. 2007.

48. However, as I show below in the chapter, and as other science studies scholars
have observed, the term "population" can and does mean "race" in actual
scientific practice (unintentionally, but no less ideologically).

49. See Bar-Sade et al. 1998.

50. Mefford et al. 1999. The mutation 943ins10 was detected in individuals residing in Ivory Coast, the Bahamas, Miami, and Washington, D.C.

51. Ibid., 576.

52. The BRCA genes are thought to increase risk for both breast and ovarian
cancer. See chapter 3.

53. Mefford et al. 1999, 576–77.

54. Gao et al. 2000b.

55. Ibid., 192.

56. Fregene and Newman 2005, 1540–41.

57. Ibid., 1541. The authors concluded: "The parallels between African American
and sub-Saharan African breast cancer suggest the possible effects of hereditary factors, and these influences may cause the younger age distribution that
is seen among these patient populations to persist" (1548). The presumption
that hereditary factors are at work is common in the breast cancer literature. Another group of researchers, reviewing previous studies of the cancer
susceptibility gene p53, wrote: "In an attempt to identify the molecular and
biological factors associated with the more aggressive behavior of breast cancer in African American women, several studies have been conducted which
report higher prevalence of somatic p53 alterations in primary breast tumours
of African American women, differences in breast density, decreased expression of isoforms of [estrogen] receptors, and alterations in other molecular
and genetic markers" (Haffty et al. 2006, 133).

58. Recurrent mutations can be, but are not necessarily, founder mutations.

59. I thank Karla Scott of St. Louis University for raising this point on a panel on
African American women and breast cancer at the annual National Commun-
ciation Association conference, San Antonio, TX, November 2006.

60. Feldman, Lewontin, and King 2003, 374.
61. Ibid.
62. Braun 2002, 167.
63. Feldman, Lewontin, and King 2003, 374.
64. For the perspectivalism argument, see Keller 1992b.
65. Epstein (2007), drawing on Duster (1990), makes this excellent point as well. Geneticists could take the population of New York City and find alleles that are more frequent than in populations defined by a different city. As Epstein points out, there just isn't any monetary or other motivation for such projects.
66. C. Clarke et al. 2003.
67. Brawley 2003b, 14.
68. Olopade et al. 2003, 238, 240. See also Panguluri et al. 1999; Kanaan et al. 2003.
69. Repeated Medline and Google searches failed to reveal any titles with the words "white" or "Caucasian" and "BRCA."
70. See de la Hoya et al. 2002 for information regarding Spanish mutations. See also German Consortium for Hereditary Breast and Ovarian Cancer 2002.
71. Although "white" is typically referred to as a race, white women are studied with much more nuance than black women, as I discuss below.
72. Table 4 in Haffty et al. 2006 (135) lists many mutations of unknown significance in white subjects.
73. John et al. 2007.
74. M. Hall et al. 2009. The groups were: Ashkenazi Jewish, Central/Eastern European, Latin American/Caribbean, Near/Middle Eastern, and Western/Northern European (2224). These categories—along with the presumably "racial" labels "African," "Asian," and "Native American"—are drawn from the test requisition form administered during a BRCA screening.
75. M. Hall et al. 2009, 2229; John et al. 2007, 2874. In both articles, the authors remind the reader that black women are more likely to be diagnosed at a young age with certain types of tumors, profiles that point to inherited susceptibility as the culprit. When genomics studies are more specific in describing subjects with African ancestry—for instance, those that employ the label "Yoruba," "Yoruba" seems to stand in for "African"—rather than complicate it (as in the case of the Hap Map project; see International HapMap Consortium 2005). And, more important, these studies are not about breast cancer.
76. John et al. 2007; Haffty et al. 2009. The neologism "young breast cancer" emerged in 2009 to denote this racialized disease category.
77. Weitzel et al. 2007, 1615.
78. Ibid., 1620.
79. Montagu 1964.
80. For an excellent analysis of the United Nations document, arguing why it did not dispense with race, see Reardon 2004.

81. Lewontin 1972.
82. For an example of this type of criticism, see F. Jackson 1998.
83. The word "penetrance" refers to the proportion of individuals of a particular genotype that express its phenotypic effect in a given environment.
84. I discuss this in greater detail in chapter 5.
85. This definition comes from N. Risch, Burchard, Ziv, and Tang 2002. By "primary," the authors presumably do not mean Africa, although they embrace the so-called out of Africa hypothesis. "Primary" appears to mean the continent reached as a result of "one or more migration events out of Africa within the last 100,000 years" (3).
86. Marks 2006.
87. Leroi 2005.
88. See, for example, Wilson et al. 2001; Romualdi et al. 2002; and Bamshad et al. 2003.
89. N. Risch, Burchard, Ziv, and Tang 2002, 2.
90. Tang et al., 2005, 274.
91. Bolnick 2008. The Tang et al. 2005 study I discuss above is one such example.
92. Feldman, Lewontin, and King 2003, 374.
93. Fujimura and Rajagopalan 2011, 22.
94. International HapMap Consortium 2005.
95. The populations selected were defined as having African, Asian and European ancestry and included the Yoruba in Nigeria, unrelated Japanese in the Tokyo, Han Chinese in Beijing, and thirty samples from Utah.
96. Nanda et al. 2005, 1932.
97. Gao et al. 2000b, 193–94. Three mutations suspected of increasing risk for breast cancer were detected, two of them novel. An additional twenty-four mutations were identified; however, at the time, none of them was thought to undermine proper gene function. It is important to note that BRCA variations are sometimes reclassified once it becomes known that they are in fact deleterious mutations.
98. Nanda et al. 2005, 1932.
99. Panguluri et al. 1999, 31.
100. These definitions come from Apple's "Dictionary," 2.1.3 (80.4), part of Mac OSX Version 10.6.7.
101. I discuss elsewhere in this chapter how and why the use of racial categories can be very dangerous medically.
102. In a similar vein, Risch and his colleagues term this the difference between variation and differentiation. In their exact words, the "greatest genetic structure that exists in the human population occurs at the racial level" (N. Risch, Burchard, Ziv, and Tang 2002, 4)—race being broadly defined as the primary continent of origin. They claim that despite relative diversity among racial

groups, individuals can nevertheless be grouped racially if one looks closely enough at genetic markers that distinguish the said groups according to evolutionary migrations out of Africa.

103. Long and Kittles 2003.

104. Stevens 2008, 324. For the language of "unique" mutations, see Gao et al. 2000a. For the research on "German" mutations, see German Consortium for Hereditary Breast and Ovarian Cancer 2002.

105. Stevens 2002.

106. Jones and Chilton 2002.

107. C. Clarke et al. 2003.

108. Krieger 2002, 612.

109. Trock 1996, 12.

110. C. Clarke et al. 2003, 215.

111. Brawley 2003a, 8.

112. For background on the field, see Steven Epstein 2007.

113. Brawley 2002, 471. For a discussion of residual confounding, see Kaufman, Cooper, and McGee 1997.

114. Institute of Medicine 2001, 8.

115. Krieger 2001, 672. For political economy approaches, see Lewontin 1991, 46; See also Levins and Lewontin 1985.

116. Krieger 2001, 668.

117. Ibid., 672.

118. Ibid., 673.

119. Ibid.

120. Most of the funding comes from the National Institutes of Health, specifically, the National Cancer Institute, the NIEHS, the National Institute on Aging, and the Office of Behavioral and Social Sciences Research.

121. Centers for Population Health and Health Disparities 2007, 12.

122. Gehlert et al. 2008, 340.

123. Centers for Population Health and Health Disparities 2007, 36.

124. Ibid., 34. See also McClintock et al. 2005.

125. I say much more about this hypothesis in chapter 5.

126. The racial stratification of chemical exposures is well documented. In one 1977 study, levels of DDT in blacks were twice those in whites, and researchers were twice as likely to find traces of lindane in black tissue samples as in white ones (Kutz, Yobs, and Strassman 1977).

127. Wolff, Britton, and Wilson 2003, 294.

128. Millikan et al. 2000.

129. McGregor et al. 1977; Russo, Tay, and Russo 1982; Trock 1996; Wolff, Britton, and Wilson 2003.

130. Krieger 1989, 210.

131. Martinot 2007, 81. Martinot discusses the example of prosecuting and jailing black mothers for drug use rather than social reform that would prevent the drug abuse in the first place.
132. Omi and Winant 1994, 185.
133. Solinger 1992, 59.
134. Chowkwanyun 2011, 261.
135. Stevens 2008.
136. Chowkwanyun, 2011 266.
137. Fields 1990.
138. Bolnick 2008, 81.
139. Fausto-Sterling 2004; Reardon 2004; Fujimura and Rajagopalan 2011.
140. M. Hall et al. 2009. However, it should be noted that the criteria for "high risk" they use are broader and more inclusive than that used by earlier BRCA researchers.
141. Epstein 2007, especially chapter 10.
142. Stevens 2003, 1075. By "disparities" I take Stevens to mean disparities in the frequency of particular genetic variants. What is more, researchers must also demonstrate whether or not statistically significant genetic variation is detectable among differently defined groups (for example, differences in geographic location) in order (1) to test the claim that "racial" differences are unique and important and (2) to prevent data mining. Stevens discusses in greater detail how her proposal could be implemented, both in this essay and in Stevens 2008.
143. Fausto-Sterling 2004, 31.

CHAPTER 5

1. I employ the term "environmental genomics" to denote the wide-ranging set of research projects among several institutions and to describe, in a general way, the intersection of environmental and genomic science. This intersection can be found in genomics proper, in epidemiology, and in public health.
2. For historical and critical approaches to public health, see the essays in Colgrove, Markowitz, and Rosner 2008.
3. It is important to note that with the passage of the Toxic Substances Control Act in 1976, a distinction was made between "old" and "new" chemicals. Most old chemicals were grandfathered into the new regulations; the EPA has had much better success in regulating new chemicals. But even then, the EPA must prove that chemicals do not pose "reasonable harm" to the public. Risk, then, is heightened for the public and lessened for corporate interests (Locke and Myers 2010).
4. Plastic that is free of *Bisphenol A* (better known as BPA), to take just one example, is a product voluntarily produced by manufacturers, thus allowing consumers more options for reducing their body burden of a chemical that

several studies suggest is dangerous and should not be ingested (although, presumably, alternatives are not tested either, giving the public a false sense of security).

5. Kuehn summarizes this: "Repeated studies have shown that people of color and low-income groups live disproportionately closer to sources of pollution and waste and have disproportionately greater exposure to toxic substances. An extensive analysis of the empirical evidence revealed that all but one of twenty-seven studies of the proximity of communities to waste facilities and toxic releases found disparities by race and income. Each of the ten studies of ambient air pollution found that communities of color and lower incomes were subject to greater amounts of air pollution than white or more affluent communities. All of the studies of exposures and health effects also documented racial and income disparities, finding, for example, higher rates of exposure to lead in black and poor children; racial and income disparities in the levels of pesticides in the blood and fat; and disproportionate exposure to high levels of toxins in fish" (1996, 118–19). Dowie provides additional specifics: "Studies completed between 1990–1995 have shown that almost 50 percent of all African-American infants tested for lead had levels higher than the CDC standard; [that, according to an EPA study in 1992,] 3 out of 4 toxic waste dumps that were not in compliance with federal regulations were located in black and Hispanic communities; that communities with hazardous waste incinerators comprise 89 percent more people of color than the national average; that at least 15 million African Americans live in communities with one or more 'uncontrolled' toxic-waste sites; that about half of all Asian Americans, Pacific Islanders, and Native Americans also live near toxic-waste sites; that more than 200 million tons of radioactive waste are located on Native American reservations; that the rate for cancers of the reproductive and sex organs among Navajo teenagers is 17 times the national rate; that every year 300,000 farm workers, mostly Hispanic, suffer from pesticide-related illnesses; that pollution-induced asthma among inner-city African Americans is much higher than the average for whites and it kills five times as many blacks as whites; and that African-American male children are three times as likely to die of asthma than white male children" (1995, 145–46).

6. Bullard 1990, 8.

7. Bullard defines "disadvantage" as "poverty, occupations below management and professional levels, low rent, and a high concentration of black residents [due to residential segregation and discriminatory housing practices]" (ibid.).

8. Ibid., 20.

9. Levins and Lewontin 1985; Krieger 2001. For a social history of epidemiological theories and practices, see Krieger 2011.

10. In the candidate gene approach, which has historically been more

time-consuming and expensive, researchers set out to find a gene whose mutations are responsible for a particular disease. This method tended to be restricted to high-penetrance alleles, such as those associated with BRCA. Recall that the BRCA researchers studied only cancer syndrome families, especially women diagnosed at a very early age, to finally isolate a section of chromosome 17 where the BRCA was thought to reside. In contrast, whole-genome sequencing to catalog polymorphisms has no such goal, although the information can be used by researchers who want to identify connections between one or more polymorphisms and disease. One example is Alzheimer's disease.

11. Single nucleotide polymorphisms are so named because variation is defined by just one nucleotide change; nucleotides make up DNA, and they are known by the abbreviations A, C, G, or T.

12. According to the project's website, its goal "is not to identify these disease-related genes directly. Rather, by identifying haplotypes, the HapMap provides a tool that can be used in what are called association studies. For these studies, researchers will compare the haplotypes in individuals with a disease to the haplotypes of a comparable group of individuals without a disease (the controls). If a particular haplotype occurs more frequently in affected individuals compared with controls, a gene influencing the disease may be located within or near that haplotype. Common diseases such as cancer, stroke, heart disease, diabetes, depression, and asthma usually result from the combined effects of a number of genetic variants and environmental factors. According to an idea known as the common disease-common variant hypothesis, the risk of contracting common diseases is influenced by genetic variants that are relatively common in populations" (International HapMap Project 2012).

13. "Tracking down chemical suspects" 2002, 10. For a representative example of Davis's inclusion of gene-environment interaction, see D. Davis et al. 1993. The most recent "State of the Evidence" report by Breast Cancer Fund includes sections on gene-environment perspectives (see Gray, Nudelman, and Engel 2010). On the NIEHS web page for the Breast Cancer and the Environment Research Program, there is a prominently placed link at the top of the page labeled "Gene-Environment Interaction" that takes the reader to a page explaining the role of inherited susceptibility to environmental pollutants.

14. Coughlin 1999, 90.

15. Established in 1997 as the Office of Genetics and Disease Prevention, this was renamed the Office of Genomics and Disease Prevention in 2003. It became the National Office of Public Health Genomics in 2006.

16. Khoury 2001.

17. Khoury 2003.

18. National Human Genome Research Institute 2011.

19. Biomarkers are forms of physical evidence that indicate a particular biological

state. In environmental science, exposure biomarkers are used to evaluate physical or chemical changes in response to a particular exposure. An example would be DNA alterations brought on from exposure to environmental toxins. For an excellent essay on the links between biomarker development and biomedicalization of disease, see Shostak 2010.

20. U.S. Department of Health and Human Services (2006). See also Schwartz and Collins 2007.

21. National Institute of Environmental Health Sciences 2008; National Institute for Environmental Health Sciences N.d.

22. See, for example, Levins and Lewontin 1985.

23. Keller 2000.

24. Schwartz and Collins (2007) seem to suggest that gene therapy is unrealistic, especially given more plausible opportunities for modifying exposures.

25. Ibid. For the discussion of how genomics can improve environmental regulations, see National Institute of Environmental Health Sciences 2000.

26. No long-term damage was inflicted on the apple industry: "Within five years, in fact, growers' profits were 50 percent higher than they had been at the time of the 60 Minutes broadcast" (Rampton and Stauber 2001, 27). Although losses were suffered, according to Rampton and Stauber, much of them were due to a glut on the market that preceded release of the National Resources Defense Council Alar study on which the 60 Minutes segment was based. For more information on the council's report, and the argument that it reported on evidence already known to the EPA using standard, accepted methods of risk assessment, see Tesh 2000.

27. Ibid., 28.

28. Any attempt by government to regulate—whether to contain occupational or public health hazards or raise the minimum wage, for example—will be resisted, as it is the responsibility of corporations and industry to maximize profits. An easy solution, of course, is for government to provide resources to help industry transition to safer alternatives. To say that we need genetic testing because regulations are too costly is like saying that because some antipoverty programs do not work, we should instead explore the relationship between poverty and individuals' genetic susceptibility.

29. Lan et al. 2004.

30. For example, in 2009 the Lan team screened for SNPs and determined that genetic variants explained differences in susceptibility to benzene hematotoxicity (toxicity in the blood; benzene is implicated in blood as well as bone marrow diseases).

31. Another term that theorists find useful in describing the impact that genomics has on public health regulation is "geneticization" (Sharp and Barrett 2000; Willett 2002; Lippman 1991; Edlin, 1987), or, the growing importance placed by doctors, public health officials and researchers, and biologists on exploring the genetic basis of disease at the expense of exploring other causes: "Overly

enthusiastic expectations regarding the benefits of genetic research for disease prevention have the potential to distort research priorities and spending for health" (Willett 2002, 695).

32. Lan et al. 2009. The benzene exposure standard employed by the Occupational Safety and Health Administration had been the litmus test for regulations requiring the removal of some chemicals from commercial use on the ground that any exposure was unacceptably dangerous. In the 1980 case *Industrial Union Department, AFL-CIO v. American Petroleum Institute*, the Supreme Court threw out that standard (448 U.S. 607 [1980]). This decision was a major victory for advocates of risk assessment premised on managing, rather than eliminating, exposures.

33. National Institute of Environmental Health Sciences n.d.

34. In the 1970s, several states passed laws requiring mandatory screening for sickle cell anemia, even though screening for other, more debilitating and lethal genetic diseases, such as Tay-Sachs, was voluntary (Duster 1990, 45). Despite the fact that sickle cell anemia poses no public health danger, children were screened as a condition of school enrollment (ibid., 39); screening was also required before getting marriage licenses and before being admitted to, or treated in, hospitals (53).

35. Ibid., 26.

36. Hubbard and Henifen 1985. The 2008 Genetic Information Nondiscrimination Act enacted, for the first time, protections against the misuse of genetic testing results. According to the Council for Responsible Genetics (2012), however, there are many exceptions to the law that need to be addressed by future legislation.

37. Below in the chapter I discuss how the testing of bodily tissues can provide the ground for a progressive, embodied, environmental health social movement.

38. Environmental breast cancer is discussed in greater detail below in the chapter.

39. Weis et al. 2005, 841; Schwartz and Collins 2007, 695.

40. Weis et al. 2005, 841.

41. In one scenario, a patient would bring his or her physician exposure data gleaned from a biosensor, and the physician would then examine biomarkers related to the exposure. Examples of treatment might include vitamin therapy, dietary changes, exercise, and other stress-reduction strategies (see, for example, Hu et al. 2002).

42. For an analysis of the privatization of public health, see the March 2001 special issue of the *American Journal of Public Health*, especially Rockhill 2001.

43. Schwartz and Collins 2007, 695.

44. Ibid. Schwartz and Collins also write that "the best opportunity to reduce risk in genetically susceptible people for the foreseeable future will not be to re-engineer their genes, but to modify their environment. The successful dietary treatment of phenylketonuria is a clear example" (2007, 695).

45. Writing about the ethical, social, and legal implications of the EGP, the bio-ethicists Sharp and Barrett write: "In part, [debates about the EGP] concern possible discriminatory uses of genetic information, but the more fundamental issue is how information on genetic risks will alter our views on personal responsibility for one's health" (2000, 280).

46. For work in critical public health, see Petersen and Lupton 1996.

47. Douglas and Wildavsky 1983.

48. Waldby 1996, 3.

49. Public health scholars, following the work of Foucault, have documented the ways in which, in an age of public health and "risk," our ways of being in the world have fundamentally shifted. Put another way, the demands of the ethical life have changed (see, for example, Petersen and Lupton 1996). I do not wholly disagree with this, although when it comes to race, I argue that it is important to study the ways in which racism results in different types of imperatives for citizen-subjects and in which lax environmental regulations exact disportionate effects on African Americans and other minority groups—with deadly results. I discuss these points further in the concluding chapter.

50. Schwartz and Collins 2007; Weis et al. 2005.

51. National Institute of Environmental Health Sciences 2008; Weis et al., 2005, 846.

52. See Steven Epstein 2007 for an excellent history and critical analysis of what he calls the "inclusion paradigm."

53. My argument here is similar to Duster's in *Backdoor to Eugenics* (1990), in which he discusses how associations between alleles and disease in socially defined racial groups become grounds for understanding such groups as members of biological races. Here I am suggesting that what happens is that disease groups are defined first by spatial location, which then becomes the ground for transforming these disease communities into racial groups. It is, then, a much more insidious process.

54. Miranda and Dolinoy 2005, 226.

55. For example, Miranda and Dolinoy (2005) assert that when researchers control for confounding variables such as income and housing, race still determines disproportionate blood lead levels. It is hard to fathom how one could control for all the non-hereditary reasons that might explain why race is correlated with elevated blood levels. Similar to the examples I discuss in chapter 4, we see a rush to proffer biological and implicitly racialized explanations of health disparities.

56. Enforcement of regulations is another matter altogether, and one does not need genomics for that. Moreover, body burden testing can be used—as it has already been—to determine whether lead regulations are working. And finally, it is worth noting that Herbert Needleman, whose use of tooth and blood

assays demonstrated adverse health effects at low levels of lead exposure, was the subject of industry-led attacks on his credibility (he was accused, but cleared, of scientific misconduct; see Rosner and Markowitz 2005). Not only does this story show that one need not demonstrate genetic susceptibility to understand the danger of lead for all children, but it also shows that scientific innovation alone does not account for the successful translation of data into policy.

57. Steingraber 1998. See Samuel Epstein 1998 and Proctor 1995 on the politics of cancer research.

58. Lewontin 1991, 46. Leone Pippard of Canadian Ecology Advocates poignantly makes this point when he asks whether it is the "lifestyle" of the beluga whale that accounts for its pollution-induced illness: "Tell me, does the St. Lawrence beluga [whale] drink too much alcohol and does the St. Lawrence beluga smoke too much and does the St. Lawrence beluga have a bad diet . . . is that why the beluga whales are ill? . . . Do you think you are somehow immune and that it is only the beluga whale that is being affected?" (quoted in Steingraber 1998, 139).

59. Proctor 1995, 4; D. Davis 2000; Kolata 2009.

60. Steingraber 1998.

61. Proctor 1995, 16–34; Steingraber 1998.

62. Gray, Nudelman, and Engel 2010. In 1993 D. Davis et al. published their "xenoestrogen" hypothesis, which posits that many synthetic chemicals act as xenoestrogens by mimicking the effect of estrogen or by blocking the hormone's normal activity in the body. The pathways of effects include increasing the production of "bad" estrogens (that is, those that increase breast cancer risk), binding to a cell's estrogen receptors and causing cell proliferation, helping tumor cells generate the blood vessels needed to grow and spread, and damaging DNA. Xenoestrogens include chlorinated compounds (for example, the pesticide DDT), PCBs, triazine herbicides, polycyclic aromatic hydrocarbons, and some pharmaceuticals (for example, DES, once given to pregnant women to prevent miscarriage and now known to be the cause of a rare vaginal cancer and other reproductive and development abnormalities in their daughters; sons of these women were also adversely affected). See also D. Davis and Bradlow 1995 for an elaboration of the xenoestrogen theory.

63. Chemicals with probable links to breast cancer include those found in most cosmetics and other personal cleaning and grooming products and pesticides currently on the market. See Evans 2006.

64. The phrase "environmental endocrine hypothesis" was coined by the historian Sheldon Krimsky (2000), and the theory behind it was popularized earlier, in *Our Stolen Future* (Colborn, Dumanoski, and Myers 1996). The hypothesis holds that "a diverse group of industrial and agricultural chemicals in contact

with humans and wildlife has the capacity to mimic or obstruct hormone function—not simply disrupting the endocrine system like foreign matter in a watchworks but fooling it into accepting new instructions that distort the normal development of the organism" Krimsky 2000, 2). The environmental endocrine hypothesis casts significant doubt on key paradigms within the field of toxicology and the US regulatory structure that it informs. Specifically, because endocrine systems are homeostatic—that is, they operate according to self-regulated feedback systems—the general understanding among toxicologists that lower doses do not result in harmful effects has been cast into doubt (23). The environmental endocrine hypothesis, therefore, challenges toxicology's inattention to the effects of low dosages on developing organisms and suggests that effects may even be "shut off" if the dose is high enough (124).

65. Estrogen is implicated in the proliferation of mammary cells, including tumor cells; it also plays a role in pregnancy-related risk because the differentiation of mammary cells, a byproduct of pregnancy, reduces risk.

66. The manufacture of PCBs, which belong to the class of organochlorines, was banned in the United States in 1977, although controversy continues over how best to deal with their lingering presence in the environment. For an investigative study about PCB removal in the Hudson River in New York State, see Cray 2001.

67. Allsopp et al. 1998.

68. Wolff et al. 1993. See also Wassermann et al. 1976; Mussalo-Rauhamaa et al. 1990; Falck et al. 1992.

69. Wolff et al. 1993, 652.

70. Hunter and Kelsey 1993, 598. When considering cancer rates, it is important to adjust for age because rates are partially explained by aging populations (that is, cancer risk increases with age).

71. Dewailly et al. 1994.

72. Dewailly, Ayotte, and Dodin 1997, 888.

73. Høyer et al. 1998.

74. Cohn, Wolff, Cirillo, and Schultz 2007.

75. See chapter 2.

76. A. Martin and Weber 2000; National Cancer Institute 2012.

77. S. Roberts et al. 1999, 785. As one team of researchers framed the issue, "it is estimated that about 5 percent of breast cancer cases are related to rare but highly penetrant genes, such as BRCA1 and BRCA. Because they are more common, however, low-penetrant cancer susceptibility genes in drug metabolism and DNA repair may contribute to a large proportion (> 90 percent) of breast cancer cases" (Hu et al. 2002, 208). It is noteworthy that of the dozens of articles on BRCA I have read, this gives by far the lowest estimate of the percentage of cases explained by the BRCA genes. The most common estimate is 5–10 percent; some estimates are higher. One essay on genetic repair limited

BRCA's contribution even more: "However, epidemiological studies have shown that mutations in these genes are relatively rare in the general population and each is likely to account for less than 1 percent of all breast cancer cases" (Angèle and Hall 2000, 168).

78. Like BRCA, p53 a tumor suppressor gene.

79. Coughlin and Piper 1999, 1023. See also Angèle and Hall 2000, 168.

80. Gray, Nudelman, and Engel 2010, 62.

81. An autosomal recessive disorder means both inherited copies of a gene must be defective for the disorder to develop.

82. Bennett notes: "AT is an inherited autosomal recessive disease characterized by cerebellar degeneration, dilated blood vessels in the eye, immune deficiencies, degeneration of the thymus, and sensitivity to ionizing radiation" (1999, 145).

83. Heterozygosity means that the two alleles of a given gene are different.

84. See, for example, Mamon et al. 2003; Angèle and Hall 2000.

85. Cited in Bernstein et al. 2002, 249. According to previous research, the cancer most likely associated with AT was breast cancer: "With a frequency of approximately 1.4 percent in the general population, as many as 14 percent of breast cancer patients could be ataxia-telangiectasia heterozygotes" (Lavin et al. 1994, 1628).

86. Ahmed and Rahman 2006. One prominent example of research establishing that risk is the Women's Environmental, Cancer, and Radiation Epidemiology study: "On the basis of early estimates that the frequency of A-T carriers (individuals with one mutated ATM gene) was 1 percent, it was suggested that A-T heterozygosity was the best candidate for a common predisposing condition and could be associated with about 5 percent of all breast cancers. This is substantially greater than the 1 percent–2 percent associated with mutations in BRCA1 and BRCA2 genes. Interest in this observation was rejuvenated when the ATM gene was cloned in 1995" (Khanna 2000, 795).

87. Duell et al. 2001, 217.

88. Hu et al. write: "Although human breast cancer risk related to DNA-repair SNPs may be small compared to cancer genes (e.g., BRCA1 and BRCA2 genes), the proportion of cancer cases attributable to them may be large because of their high frequency in the general population" (2002, 213).

89. Ibid., 209.

90. Smith et al. 2003, 1203.

91. Medina et al. 2002, 182.

92. See, for example, the work of Breast Cancer Fund. Their "State of the Evidence" reports provide insightful overviews of the science behind the environmental breast cancer thesis (see, for example, Gray, Nudelman, and Engel 2010).

93. The risk of cancer developing in the healthy breasts of patients was an early motivating factor behind this research.

94. Khanna 2000, 800. See Jameel, Rao, Cawkwell, and. Drew 2004 for a review of the literature on radiotherapy and radioresistance.

95. Hu et al. 2002, 208. The genes in question are the XRCC1, XRCC3, and APE1 genes.

96. Medina et al. (2002) employ the term "genomic instability" to describe this inborn condition.

97. Hu et al. 2002, 209.

98. Ibid.

99. Ibid., 208.

100. Ibid., 213.

101. McClintock et al. 2005.

102. Hu et al. 2002, 213. See also Smith et al. 2008.

103. Guillemette, Millikan, Newman, and Housman 2000. The Carolina Breast Cancer Study, started in 1993 (and now recruiting subjects for phase 3) examines disparities between African Americans and whites. Women are interviewed, their medical records assessed, and they provide a DNA sample. The UGT1A1 is not a repair gene, but it produces an enzyme that modifies the metabolism of estrogen—typically endogenous estrogen, but perhaps also estrogenic chemicals like those found in plastics and pesticides. Thus, its actual role in breast cancer susceptibility is different from the role of a repair gene, but UGT1A1 has captured researchers' interest for reasons similar to those of the repair genes: to find susceptibility alleles that are low penetrant, yet widely dispersed (more prevalent) in specific populations.

104. See Kahn 2004 for a devastating critique of the rationale behind the Food and Drug Administration's approval to market the heart medication Bidil to African Americans.

105. UGT genes are involved in the making of enzymes that are in turn involved in metabolism (for example, metabolism of carcinogens).

106. Guillemette, Millikan, Newman, and Housman 2000, 955.

107. Other factors, such as gender, certainly play a role in environmental justice politics; for now, however, I am focusing on race and class. Below in the chapter I address the topic of gender justice in the context of environmental breast cancer activism.

108. See Dowie 1995, 142.

109. See ibid.

110. See ibid., 143.

111. Markowitz and Rosner 2002. Louisiana's relationship with the petrochemicalindustry dates back to the early twentieth century; it is home to oil wells, refineries, and offshore drilling (ibid., 234–40): "By the mid-1950s, chemicals

and chemical products ranked first in the value of manufactured products in Louisiana" (239).

112. Ibid., 237–39.

113. Ibid., 239–40. By the mid-1990s, Louisiana's per person releases of toxic substances were three times that of Texas, which at the time ranked second in the country (272).

114. Quoted in Bullard 1990, 34.

115. See Dowie 1995.

116. Quoted in Dowie 1995, 142–43.

117. Ibid., 143.

118. According to Bullard (1990, 33), industry has taken advantage of the fact that it is harder for residents to mobilize when the majority are renters.

119. Markowitz and Rosner 2002, 239.

120. Quoted in Dowie 1995, 143.

121. West and Melancon 2002.

122. Markowitz and Rosner 2002, 273–75.

123. Ibid., 265.

124. West and Melancon 2002.

125. Ibid.

126. Dowie 1995. The EPA defended the proposal even though the geology of the area virtually assured that PCBs would leak into the groundwater (Bullard 1990, 36).

127. Bullard 1990, 35–38.

128. Ibid., 17.

129. In 1994 President Bill Clinton signed Executive Order 12898, which made environmental justice part of the mission of all federal agencies.

130. Markowitz and Rosner 2002, 272. For another history of the Shintech controversy, see J. Roberts and Toffolon-Weiss 2001. PVC is short for polyvinyl chloride, which is a component of many plastics. It is made with vinyl chloride (a gas), the use of which is at the center of many lawsuits due to its connection with cancers, including those of the brain and liver.

131. Convent activists had plenty of precedent on which to build their arguments. Because of tax breaks and other incentives, the state of Louisiana spent $10.7 million per job created by a Dow Chemical plant in Plaquemine. The *New York Times* reported that in 2000 Louisiana had the second highest poverty rate of any state and that the gap between rich and poor was the largest in nation (cited in Markowitz and Rosner 2002, 273). High-tech jobs, more often than not, have been reserved for white professionals from outside a given parish or even outside the state. In the town of St. Gabriel, for instance, one survey found that local residents held just 164 out of 1,878 permanent jobs (274). And the impact of tax breaks and other incentives on public services like schools has ensured that the workforce

remains unskilled (West and Melancon 2002). Bullard notes: "Even with the economic transformation, many of the region's old problems that were related to underdevelopment (e.g., poor education, large concentrations of unskilled labor, low wages, high unemployment, etc.) went unabated" (1990, 32–33).

132. Markowitz and Rosner 2002, 272–73.

133. Ibid., 277–78. For an excellent study of toxic tours, see Pezzullo 2007.

134. Markowitz and Rosner 2002, 284.

135. For the language of chemical assault, see Bullard 2002.

136. Sandler and Pezzullo 2007; P. Brown 2007.

137. After deciding not to pursue its plans for construction in Convent, Shintech won approval to build a plant on a site across the river. Shintech avoided charges of environmental racism because no one lived at the new site—the entire community had been relocated by Dow in 1991 (Markowitz and Rosner 2002, 285).

138. One of the main lines of attack involved characterizing activists as dupes and puppets of the local and national organizations with which they were allied, including the Louisiana Environmental Action Network, Greenpeace, and Tulane Law School's Environmental Law Clinic. The president of the Louisiana Association of Business and Industry publicly claimed that the civil rights argument was just a misguided attempt to revitalize a defunct movement: "Environmental justice is largely the creation of activists who turned to civil rights when environmental doom-saying no longer yielded results" (quoted in J. Roberts and Toffolon-Weiss 2001, 192). The governor weighed in eventually, attacking the clinic publicly; the secretary of the Louisiana Department of Economic Development attacked and attempted to discredit law clinic workers as elitists whose "agenda" undermined the chances of poor residents to obtain well-paying and secure employment (ibid., 113).

139. La. Sup. Ct. R. XX, full text available at http://www.lasc.org/rules/html/xx499.htm.

140. Quoted in J. Roberts and Toffolon-Weiss 2001, 202. Widespread resistance led to the watering down of these restrictions in 1999 (Markowitz and Rosner 2002, 281–82), but it is nevertheless the case that few of the clinic's clients at the time would have been able to meet the new stipulations (J. Roberts and Toffolon-Weiss 2001, 202).

141. Quoted in J. Roberts and Toffolon-Weiss 2001, 201.

142. Proctor 1995.

143. Wagner 1995.

144. See P. Brown 1987 for a study of experiential knowledge and environmental health activism. Tesh complicates this in an interesting way, noting that a false dichotomy exists, in which activists are presumed to "feel" risks, while experts are presumed to "know" them (2000, 85).

145. They thus employ the concept "embodiment" in different ways than does Klawiter (2008).

146. Biomonitoring, is, more or less, in the early stages of becoming a widespread, common practice. The arguments I make in this section are, in some ways, necessarily speculative.

147. Beck 1992. The exception, of course, is the people living in areas I described in the previous section on environmental racism. For these individuals, the noxious effects of pollution are experienced on a daily basis, assaulting the eyes, ears, lungs, and skin. Moreover, the effects of climate change (flooding of coastal areas, droughts, and so forth), also a major environmental issue, are becoming more and more palpable.

148. Ibid., 23.

149. Ibid., 28.

150. Ohmann 1996, especially chapters 1–3.

151. See chapter 2 for discussion of the professional-managerial class along these lines.

152. For a critique of the deficit model, see Wynne 1993.

153. Risk experts assume that "public panic" often results when risk is "unknown, dreaded, involuntary, hidden, and uncontrollable" (Tesh 2000, 90). The problem with the research on which this claim is based is that although grass-roots groups do in fact critique reductionist notions of risk, they also act when risk is clearly quantifiable, and they employ scientific standards of measurement in the process. An example is Alar (92–93). Moreover, risk assessment and risk communication experts assume that only lay activists inject values into deliberative decision making. Of course, in this chapter, I am raising the question as to whether it is acceptable to demand action on ethical grounds alone.

154. It is by no means clear whether the progressive politics I envision will be realized. I simply wish to raise some theoretical possibilities regarding different body politics from that which genomics allows (I explore this further in the book's concluding chapter). Nevertheless, we must be somewhat wary of biomonitoring in light of the body discourses it may both draw from and contribute to. Gender politics, for example, can influence both the decision to conduct biomonitoring and how the information will be used (Morello-Forsch 1997; Casper and Moore 2009). Furthermore, some environmental justice activists are understandably concerned that biomonitoring favors a narrow focus on individual diseases at the expense of more comprehensive health reform that takes into consideration important social and economic factors (Bhatia et al. 2004). Vulnerable groups who are already socially ostracized risk being further stigmatized once biomonitoring results are made public. Put another way, the bodies comprising what sociologists have termed "embodied social movements" (Klawiter 2008, P. Brown, Morello-Frosch, Stephen Zavestoski, and the Contested Illnesses Research Group 2012) are

only intelligible to researchers and policymakers because of the bodily norms that ground our interpretive frameworks. Bodies are never abstract material objects—they are culturally gendered, sexed, and raced and, as such, are value laden and normative. Science and technology studies scholars working in the area of HIV/AIDS have shown that blood is never value free: its meaning is inflected with notions of risk and pathology attributed to the bodies from which it is drawn (Waldby and Mitchell 2006). Along these lines, Phaedra Pezzullo discussed the Environmental Working Group's discourse of the "pre-polluted" from a feminist perspective during a "Spotlight Scholar" talk she gave at the University of Georgia in April 2011.

155. Breast Cancer Fund and Commonweal, in a jointly produced document on the California bill, put it this way: "Just as individuals have a right to informed consent before undergoing medical procedures, we have a right to know what industrial chemicals are in our bodies that could be harmful to our health" (n.d., 1). They continued: "The presence of hundreds of industrial chemicals in the body does not mean it is 'normal.' Californians have not given permission to put untested chemicals in their bodies and we want to know more about what harm these chemicals may cause, what products contain these chemicals, and how people become exposed to them" (2). The authors of a Commonweal study observed: "When you hear the word 'pollution,' you think of pol-luted air, polluted water, polluted soil" (Patton and Baltz 2005, 3). Without a biomonitoring project, they argued, "Californians will continue to be part of a vast chemistry experiment" (8).

156. According to Davis Baltz of Commonweal, biomonitoring "is helping to create a revolution in our understanding of the links between exposure to chemicals and disease" (Patton and Baltz 2005). The cosponsor of the bill that estab-lished California's biomonitoring program, Deborah Ortiz, claimed: "This law will help California create a strong, science-based program to establish links to chronic conditions, reduce exposure to toxic chemicals, and better protect our health and our environment" (Breast Cancer Fund 2006).

157. For excellent treatments of the environmental breast cancer movement and the science behind it, see Ley 2009 and McCormick 2010.

158. The first proposal for a statewide biomonitoring project in California would have made breast-milk testing the primary method of documenting body burden.

159. Steingraber 2001; Jackson 2000; Allsopp et al. 1998.

160. Jackson 2000, 7.

161. As Ley (2009) observes, activists have expanded their agenda to include chil-dren's health, largely as a result of data concerning breast-milk contamination.

162. Steingraber 2001. For a multifaceted study of contaminated breast milk, see Boswell-Penc 2006.

163. See Steven Epstein 2007, chapter 11, for an excellent overview of the many

problems with sex differences research. Epstein points out, for example, that although other bodily attributes, such as race, are controversial and have sparked considerable debate, gender is not controversial and in fact is an increasingly popular and accepted variable in health disparities research. This explains the significance of the different interpretation of gender offered by environmental breast cancer activists.

164. For details about the study, see Swirsky 1990. Long Island has been a site of intense organizing around the environmental theory of breast cancer for almost twenty years. This is largely due to the high rates of breast cancer there, particularly in Nassau and Suffolk Counties. In the mid-1980s, New York health officials first reported that breast cancer rates were higher in parts of Long Island than they were in the rest of the state. Between 1981 and 1985, age-adjusted breast cancer incidence for Nassau County increased from 104.4 per 100,000 females to 117.8 (a 13 percent change). Age-adjusted mortality rates decreased by 13 percent. In Suffolk County, incidence increased from 92.3 to 113.6 over the same period (a 23 percent change). Mortality decreased by 7 percent. Between 1986 and 1990, 98.9 of every 100,000 women in New York state were diagnosed with breast cancer. On Long Island, the rate was 114.7 in Nassau County and 108.6 in Suffolk County. Between 1991 and 1995, the rates were 101.1 for the state, 118.2 for Nassau, and 110.7 for Suffolk (Goldstein 1999, 42; National Cancer Institute and National Institute of Environmental Health Sciences 2000, 2). As of 1998, the incidence rate per 100,000 women was 117.1 in Nassau and 110.6 in Suffolk, while it was 101.5 for all of New York State (E. Miller 1998). These are particularly high numbers, given the fact that breast cancer rates already tend to be higher in New York than in the rest of the country. Research has also revealed that variations on Long Island can be extreme. For example, in the North Fork region, the breast cancer rate is 56.6 cases per 100,000. In the South Fork region, however, it is a striking 90.3 cases per 100,000, and in Riverhead it is 70.9 (Gearty 1998). High mortality rates have also served as a catalyst for organizing. As of 1990, mortality from breast cancer in Nassau and Suffolk Counties is 27 percent higher than the US rate, and diagnoses of metastasized breast cancer are two to three times the national average (Jenks 1994, 88). Environmental breast cancer research has continued since them, the most recent incarnation being the Breast Cancer and the Environment Research Program begun in 2009 and jointly supported by the NIEHS and the National Cancer Institute.

165. This is not a reference to BRCA mutations thought to be more prevalent among Ashkenazi Jews. I address the "religion" risk factor in chapter 4.

166. "One in nine" is a reference to the lifetime risk for breast cancer for US women. They now have a one in eight chance of getting breast cancer during their lifetimes.

167. Balaban revealed in her interview with me in September 2000 that her breast

cancer activism was in many ways an expected outcome of her early activism in other areas. See Ley (2009) and Klawiter (2008) for more on the connections between antiwar, feminist, and breast cancer activism.

168. Swirsky 1990.

169. Goldstein 1999, 46.

170. Quoted in ibid.

171. Ibid.

172. National Cancer Institute and National Institute of Environmental Health Sciences 2000, Appendix A.

173. Quoted in Jenks 1994, 88.

174. Quoted in Goldstein 1999, 43.

175. Steven Epstein 1996.

176. Popular epidemiology is a method in which laypersons gather epidemiological data and also direct and marshal the knowledge and resources of experts in order to understand the epidemiology of disease (P. Brown 1987, 78). According to the sociologist Phil Brown, "without popular participation, it would often be impossible to carry out much of the research needed to document health hazards. Science is also limited in its conceptualization of what are problems and how they should be studied and addressed" (81–82). Indeed, popular epidemiology "emphasizes basic social structural factors, involves social movements, and challenges certain basic assumptions of traditional epidemiology" (78).

177. Tesh 2000; P. Brown 1987.

178. D. Davis and Bradlow 1995, 169; Krimsky 2000, chapter 2.

179. The precautionary principle has been adopted in other countries and by international treaty negotiators (for a brief overview of the philosophy of precaution, see Raffensperger and Tickner 1999; for a history of its use, see Jordan and O'Riordan 1999). For example, in 1998, the European Parliament overwhelmingly approved the gradual removal of hormone-disrupting chemicals from the European market (Krimsky 2000, 90). Today, the European Union generally follows the precautionary principle with regard to chemical contaminants. In the United States, some precedent exists for similar action, but that precedent is limited. The National Environmental Policy Act (which applies only to the use of federal resources) in effect calls for precaution to be the guiding norm (Geiser 1999). It should also be noted that biomonitoring can be employed to determine whether regulations are working. For example, when lead regulations were enacted, body burden data provided evidence of their effectiveness, as lead levels in blood decreased significantly (Centers for Disease Control and Prevention n.d.).

180. Klawiter 2008, chapter 7.

181. See, for example, Krieger 1994a; Hunter et al. 1997; Stellman et al. 2000. A recently released National Research Council report (2012) seemed to cast

doubt on the environmental breast cancer theory. For criticism of the report, see Breast Cancer Action 2011.

182. Sternberg 1994; Krieger 1994b, 1094.

CHAPTER 6

1. Rose 2006, 119.
2. Butler 1993.
3. See Sherwin 1992.
4. In scholarship that assesses attitudes and decision making about oophorectomy, for example, there is some evidence that women do take into consideration the negative effects of the surgery, and therefore qualify genetic risk in particular ways; see Hallowell et al. 2001. Other studies have documented that women are likely to underestimate risk from oophorectomy and overestimate risk from cancer, thus showing the need to critically engage current guidelines as well as the content of genetic counseling sessions; see Fry, Rush, Busby-Earle, and Cull 2001.
5. For a thorough history and critical examination of biomedicalization, see A. Clarke et al. 2010.
6. The participation rate by African American women, for example, is exceedingly low, as I note in chapter 4. Furthermore, assessments of attitudes about surgery can and do, without reflection, implicitly condemn women for refusing it; see, for example, S. Miller et al. 2005. Future research might examine resistance as a form of thanatopolitics; see Murray 2006; Rose 2006.
7. For scholarship on biosociality, see Rabinow 1992; Rose 2006; Klawiter 2008.
8. Shildrick 2004, 163.
9. Rose 2006, 25–26.
10. Some African American women do inherit BRCA mutations, but that does not justify using race as a category for offering the test. Personal or family history of breast and ovarian cancer, not race, should be used to determine which African American women should be tested.
11. Klawiter 2008; Duster 1990.
12. Stevens 2008.
13. Lewontin 1991, 12–13.
14. See Murray 2007 for a reconsideration of bioethics in terms of a nongeneticized care of the self, a rethinking that, he says, allows for due consideration of care and love for others.

BIBLIOGRAPHY

ACOG [American Congress of Obstetricians and Gynecologists]. 2009. "Routine screening for hereditary breast and ovarian cancer recommended." http://www.acog.org/About_ACOG/News_Room/News_Releases/2009/Routine_Screening_for_Hereditary_Breast_and_Ovarian_Cancer_Recommended (accessed October 22, 2012).

ACOG [American Congress of Obstetricians and Gynecologists] Committee on Practice Bulletins. 1999. "ACOG Practice Bulletin No. 7: Prophylactic oophorectomy." *International Journal of Gynecology & Obstetrics* 67 (3): 193–99.

———. 2008. "ACOG Practice Bulletin No. 89: Elective and risk-reducing salpingo oophrectomy." *Obstetrics and Gynecology* 111 (1): 231–41.

Ahmed, M., and N. Rahman. 2006. "ATM and breast cancer susceptibility." *Oncogene* 25:5906–11.

Aliyu, Muktar H., et al. 2005. "Trends in birth across high-parity groups by race/ethnicity and maternal age." *Journal of the National Medical Association* 6 (June): 799–804.

Allsopp, Michelle, Ruth Stringer, and Paul Johnston. 2008. *Unseen Poisons: Levels of Organochlorine Chemicals in Human Tissues.* http://archive.greenpeace.org/toxics/reports/unseenpo.pdf (Accessed November 27, 2012).

Amend, Kandace, David Hicks, and Christine B. Ambrosone. 2006. "Breast cancer in African-American women: Differences in tumor biology from European-American women." *Cancer Research* 66 (September 1): 8327–30.

American Cancer Society. 2005. "Cancer facts and figures 2005." On file with author.

———. 2012. "Cancer facts and figures 2012." http://www.cancer.org/Cancer/OvarianCancer/DetailedGuide/ovarian-cancer-key-statistics (accessed October 22, 2012).

Angèle, Sandra, and Janet Hall. 2000. "The ATM gene and breast cancer: Is it really a risk factor?" *Mutation Research* 462 (2–3): 167–78.

Angier, Natalie. 1994. "Breast cancer gene isn't making screening easy." *New York Times*, November 30.

Armstrong, Katrina, et al. 2004. "Hormone replacement therapy and life expectancy after prophylactic oophorectomy in women with BRCA1/2 mutations: A decision analysis." *Journal of Clinical Oncology* 22 (6): 1045–54.

———. 2005. "Racial differences in the use of BRCA1/2 testing among women with a family history of breast or ovarian cancer." *Journal of the American Medical Association* 293 (14): 1729–36.

"Ashkenazi Jewish women linked to mutated breast cancer." 1995. *Buffalo News*, August 30.

Ayala, Francisco J. 1987. "Two frontiers of human biology: What the sequence won't tell us." *Issues in Science and Technology* 3:51–56.

Baltimore, David. 1987. "Genome sequencing: A small-science approach." *Issues in Science and Technology* 3:48–50.

Bamshad, Michael J., et al. 2003. "Human population genetic structure and inference of group membership." *American Journal of Human Genetics* 72 (3): 578–89.

Barad, Karen. 2003. "Posthumanist performativity: Toward an understanding of how matter comes to matter." *Signs* 28 (3): 801–31.

Barker-Benfield, G. J. 1976. *The Horrors of the Half-Known Life: Male Attitudes toward Women and Sexuality in Nineteenth-Century America*. New York: Harper and Row.

Barrett, Michèle. 1997. "Capitalism and women's liberation." In *The Second Wave: A Reader in Feminist Theory*, edited by Linda Nicholson, 123–30. New York: Routledge.

Bar-Sade, Revital Bruchim, et al. 1998. "The 185delAG BRCA1 mutation originated before the dispersion of Jews in the Diaspora and is not limited to Ashkenazim." *Human Molecular Genetics* 7(5): 801–5.

Batt, Sharon. 1994. *Patient No More: The Politics of Breast Cancer*. Charlottetown, PE: Gynergy.

Beatty, John. 1991. "Genetics in the atomic age: The atomic bomb casualty commission, 1947–1956." In *The Expansion of American Biology*, edited by Keith Benson, Jane Majenschein, and Ronald Rainger, 284–324. New Brunswick: Rutgers University Press.

Beck, Ulrich. 1992. *Risk Society: Towards a New Modernity*. London: Sage.

Bennett, L. Michelle. 1999. "Breast cancer: Genetic predisposition and exposure to radiation." *Molecular Carcinogenesis* 26 (3): 143–49.

Bernstein, Jonine L., et al. 2002. "Workshop on the epidemiology of the ATM gene: Impact on breast cancer research and treatment, present status and future focus, Lillehammer, Norway. 29 June 2002." *Breast Cancer Research* 4 (6): 249–52.

Bhatia, Rajiv, et al. 2004. "Biomonitoring: What communities must know." *Race, Poverty, and the Environment* 11 (2). http://www.urbanhabitat.org/node/172 (accessed November 2, 2012).

Bishop, Joseph, and Michael Waldholz. 1990. *Genome: The Story of the Most Astonishing Scientific Adventure of Our Time—The Attempt to Map All the Genes in the Human Body*. New York: Simon and Schuster.

Blair, William. 1995. "Geneticist tells of fall-out harm." *New York Times*, April 26.

Bolnick, Deborah A. 2008. "Individual ancestry inference and the reification of race as biological phenomenon." In *Revisiting Race in a Genomic Age*, edited by

Barbara A. Koenig, Sandra Soo-Jin Lee, and Sarah S. Richardson, 70–85. New Brunswick: Rutgers University Press.

Boswell-Penc, Maia. 2006. *Tainted Milk: Breastmilk, Feminisms, and the Politics of Environmental Degradation*. Albany: State University of New York Press.

Bourne, Thomas H., et al. 1991. "Ultrasound screening for familial ovarian cancer." *Gynecologic Oncology* 43 (2): 92–97.

Braun, Lundy. 2002. "Race, ethnicity, and health. Can genetics explain disparities?" *Perspectives in Biology and Medicine* 45 (2): 159–75.

Brawley, Otis. 2002. "Disaggregating the effects of race and poverty on breast cancer outcomes." *Journal of the National Cancer Institute* 94 (7): 471–73.

———. 2003a. "Cancer and health disparities." *Cancer and Metastasis Reviews* 22 (1): 7–9.

———. 2003b. "Population categorization and cancer statistics." *Cancer and Metastasis Reviews* 22 (1): 11–19.

Breast Cancer Action. 2011. "Breast Cancer Action says IOM report misses important opportunities to turn the tide on the epidemic." December 11. http://bcaction.org/2011/12/07/breast-cancer-action-says-iom-report-misses-important-opportunities-to-turn-the-tide-on-the-epidemic/ (accessed November 9, 2012).

Breast Cancer Fund. 2006. "New California environmental health program will measure pollution in people."

——— and Commonweal. N.d. "Biomonitoring and Senate Bill 1379: Myths vs. facts." www.ncel.net/articles/biomonitoring.Myths.Facts.final.doc (accessed November 9, 2012).

"Breast cancer mythology on Long Island." 2002. *New York Times*, August 31.

"Breast cancer risk likely higher for lesbians." 1998. *Arizona Republic*, August 29.

Broder, Michael S., David E. Kanouse, Brian S.Mittman, and Steven J. Bernstein. 2000. "The appropriateness of recommendations for hysterectomy. " *Obstetrics and Gynecology* 95 (2): 199–205.

Brooks, Jamie L., and Meredith L. King. 2008. "Geneticizing disease: Implications for racial health disparities." *Center for American Progress*. January. http://www.americanprogress.org/issues/2008/01/geneticizing_disease.html (accessed October 23, 2012).

Brown, Michael K., Martin Carnoy, Elliot Currie, and Troy Duster. 2003. *White-Washing Race: The Myth of the Color-Blind Society*. Berkeley: University of California Press.

Brown, Phil. 1987. "Popular epidemiology: Community response to toxic waste-induced disease in Woburn, MA." *Science, Technology, and Human Values* 12 (Summer–Fall): 78–85.

———. 2007. *Toxic Exposures: Contested Illnesses and the Environmental Health Movement*. New York: Columbia University Press.

———, Rachel Morello-Frosch, Stephen Zavestoski, and the Contested Illnesses

Research Group, eds. 2012. *Contested Illnesses: Citizens, Science, and Health Social Movements*. Berkeley: University of California Press.

Bullard, Robert D. 1990. *Dumping in Dixie: Race, Class, and Environmental Quality*. Boulder, CO: Westview.

———. 2002. "Environmental justice in the United States: Threats to quality of life." http://www.ejrc.cau.edu/summit2/EJSummitIIFactSheet.pdf (accessed November 9, 2012).

Burke, Wylie, et al. 1997. "Recommendations for follow-up care of individuals with an inherited predisposition to cancer." *Journal of the American Medical Association* 277 (12): 997–1003.

Butler, Judith. 1990. *Gender Trouble: Feminism and the Subversion of Identity*. New York: Routledge.

———. 1993. *Bodies That Matter: On the Discursive Limits of "Sex."* New York: Routledge.

Buys, Saundra S., et al. 2005. "Ovarian cancer screening in the Prostate, Lung, Colorectal and Ovarian (PLCO) cancer screening trial: Findings from the initial screen of a randomized trial." *American Journal of Obstetrics and Gynecology* 193 (5): 1630–9.

"Cancer danger for lesbians." 1993. *London Independent*, February 5.

Canguilhem, Georges. 1989 [1966]. *The Normal and the Pathological*. Translated by Carolyn R. Fawsett, in collaboration with Robert S. Cohen. New York: Zone.

Carey, Lisa A., et al. 2006. "Race, breast cancer subtypes, and survival in the Carolina Breast Cancer Study." *Journal of the American Medical Association* 295 (21): 2492–502.

Carter-Pokras, Olivia, and Claudia Baquet. 2002. "What is a 'health disparity'?" *Public Health Reports* 117 (5): 426–34.

Casamayou, Maureen Hogan. 2001. *The Politics of Breast Cancer*. Washington: Georgetown University Press.

Casper, Monica J., and Lisa Jean Moore. 2009. *Missing Bodies: The Politics of Visibility*. New York: New York University Press.

Cass, Ilana, et al. 2003. "Improved survival in women with BRCA-associated ovarian carcinoma." *Cancer* 97 (9): 2187–95.

Centers for Disease Control and Prevention. 2006. "Ovarian cancer: Facing the challenge." 2004/2005 Fact Sheet, on file with author.

———. 2012a. "Leading causes of death in females." http://www.cdc.gov/women/lcod/index.htm (accessed November 9, 2012).

———. 2012b. "Ovarian cancer statistics." http://www.cdc.gov/cancer/ovarian/statistics/ (accessed November 9, 2012).

———. N.d. "Making a difference." http://www.cdc.gov/biomonitoring/flash/presentation_popup.html (accessed November 9, 2012).

Centers for Population Health and Health Disparities. 2007. "Cells to society: Overcoming health disparities." November. http://cancercontrol.cancer.gov/

populationhealthcenters/cphhd/documents/CPHHD_report.pdf (accessed July 11, 2012).

Chafe, William H. 1999. *The Unfinished Journey: America since World War II.* New York: Oxford University Press.

Chen, Yumay, et al. 1995. "Aberrant subcellular localization of BRCA1 in breast cancer." *Science* 270 (5237): 789–91.

Chlebowski, Rowan T., et al. 2009. "Breast cancer after use of estrogen plus progestin in postmenopausal women." *New England Journal of Medicine* 360 (6): 573–86.

Chowkwanyun, Merlin. 2011. "The strange disappearance of history from racial health disparities research." *Du Bois Review* 8 (1): 253–70.

Cimons, Marlene. "Trying to map elusive N.Y. cancer source." 1999. *Los Angeles Times*, October 18.

Clarke, Adele E., et al. 2010. *Biomedicalization: Technoscience, Health, and Illness in the U.S.* Durham: Duke University Press.

Clarke, Christina A., et al. 2003. "Existing data on breast cancer in African-American women: What we know and what we need to know." *Cancer* 97 (1 Supplement): 211–21.

Clinton, William Jefferson, Tony Blair, Francis Collins, and Craig Venter. 2000. "Remarks by the president, Prime Minister Tony Blair of England, Dr. Francis Collins, director of the National Human Genome Research Institute, and Dr. Craig Venter, president and chief scientific officer, Celera Genomics Corporation, on the completion of the first survey of the entire Human Genome Project." Washington: White House, Office of the Press Secretary, June 26. http://www.genome.gov/10001356 (accessed October 22, 2012).

Cohn, Barbara A., Mary S. Wolff, Piera M. Cirillo, and Robert I. Schultz. 2007. "DDT and breast cancer in young women: New data on the significance of age at exposure." *Environmental Health Perspectives* 115 (10): 1406–14.

Colborn, Theo, Dianne Dumanoski, and John Myers. 1996. *Our Stolen Future: Are We Threatening Our Fertility, Intelligence, and Survival? A Scientific Detective Story.* New York: Penguin.

Colditz, Graham A., et al. 1987. "Menopause and the risk of coronary heart disease in women." *New England Journal of Medicine* 316 (18): 1105–10.

Colgan, Terence J., et al. 2001. "Occult carcinoma in prophylactic oophorectomy specimens: Prevalence and association with BRCA germline mutation status." *American Journal of Surgical Pathology* 25 (10): 1283–89.

Colgrove, Gerald, Gerald Markowitz, and David Rosner. 2008. *The Contested Boundaries of American Public Health.* New Brunswick: Rutgers University Press.

Collins, Francis S., Eric D. Green, Alan E. Guttmacher, and Mark S. Guyer. 2003. "A vision for the future of genomics research: A blueprint for the genomic era." *Nature* 422 (April 24): 1–13.

Collins, Patricia Hill. 2000. *Black Feminist Thought: Knowledge, Consciousness, and the Politics of Empowerment*. New York: Routledge.

———. 2004. *Black Sexual Politics: African Americans, Gender, and the New Racism*. New York: Routledge.

Commoner, Barry. 2002. "Unraveling the DNA myth: The spurious foundation of genetic engineering." *Harpers*, February, 39–47.

Commonweal. 2005. "Study finds prominent Californians contain hazardous chemicals common in consumer products." Press release, August 30. http://www.commonweal.org/programs/brc/Taking_It_All_In.html (accessed November 9, 2012).

Condit, Celeste M. 1999a. "The materiality of coding: On rhetoric, genetics, and the matter of life." In *Rhetorical Bodies*, edited by Jack Selzer and Sharon Crowley, 326–56. Madison: University of Wisconsin Press.

———. 1999b. *The Meanings of the Gene: Public Debates about Human Heredity*. Madison: University of Wisconsin Press.

Cook-Deegan, Robert. 1994. *The Gene Wars: Science, Politics and the Human Genome*. New York: Norton.

Cooper, Melinda. 2008. *Life as Surplus: Biotechnology and Capitalism in the Neoliberal Era*. Seattle: University of Washington Press.

Corea, Gene. 1985. *The Mother Machine: Reproductive Technologies from Artificial Insemination to Artificial Wombs*. New York: Harper and Row.

Coughlin, Steven S. 1999. "The intersection of genetics, public health, and preventive medicine." *American Journal of Preventive Medicine* 16 (2): 89–90.

——— and Margaret Piper. 1999. "Genetic polymorphisms and breast cancer risk: A review." *Cancer Epidemiology, Biomarkers, and Prevention* 8 (11): 1023–32.

Council for Responsible Genetics. 2012. "Genetic testing, privacy and discrimination." http://www.councilforresponsiblegenetics.org/Projects/PastProject.aspx?projectId=1 (accessed November 9, 2012).

Couto, Richard A. 1986. "Failing health and new prescriptions: Community-based approaches to environmental risks." In *Current Health Policy Issues and Alternatives: An Applied Social Science Perspective*, edited by Carole E. Hill, 53–70. Athens: University of Georgia Press.

Cowley, Geoffrey. 1993. "Family matters: The hunt for a breast cancer gene." *Newsweek*, December 6, 46–52.

Cray, Charlie. 2001. "Toxics on the Hudson." *Multinational Monitor* 22 (7–8): 9–18.

Crick, Francis. 1958. "On protein synthesis." *Symposia of the Society for Experimental Biology* 12:138–63.

———. 1970. "Central dogma of molecular biology." *Nature* 227 (August 8): 561–63.

Dally, Ann. 1991. *Women under the Knife: A History of Surgery*. New York: Routledge.

Danbom, David. 1987. *"The World of Hope": Progressives and the Struggle for an Ethical Public Life*. Philadelphia: Temple University Press.

Dandolu, Vani, and Enrique Hernandez. 2005. "Ovarian conservation at the time of hysterectomy for benign disease." *Obstetrics and Gynecology* 106 (5): 1106.

Davies, Kevin, and Michael White. 1996. *Breakthrough: The Race to Find the Breast Cancer Gene.* New York: Wiley.

Davis, Angela Y. 1981. *Women, Race & Class.* New York: Vintage.

Davis, Devra Lee. 2000. "Most cancer is made, not born." *San Francisco Chronicle,* August 10.

——— and H. Leon Bradlow. 1995. "Can environmental estrogens cause breast cancer?" *Scientific American* 273 (4): 166–72.

Davis, Devra Lee, et al. 1993. "Medical hypothesis: xenoestrogens as preventable causes of breast cancer." *Environmental Health Perspectives* 101 (5): 372–77.

Davis, Joel. 1990. *Mapping the Code: The Human Genome Project and the Choices of Modern Science.* New York: Wiley.

Davy, M., and M. K. Oehler. 2006. "Does retention of the ovaries improve long-term survival after hysterectomy? A gynecological oncological perspective." *Climacteric* 9 (3): 167–68.

De la Hoya, M., et al. 2002. "Association between BRCA1 and BRCA2 mutations and cancer phenotype in Spanish breast/ovarian cancer families: Implications for genetic testing." *International Journal of Cancer* 97 (4): 466–71.

Delgado, Richard, and Jean Stefanic. 2000. *Critical Race Theory: The Cutting Edge.* Philadelphia: Temple University Press.

———. 2001. *Critical Race Theory: An Introduction.* New York: New York University Press.

Dewailly, Eric, Pierre Ayotte, and Sylvie Dodin. 1997. "Could the rising levels of estrogen receptor in breast cancer be due to estrogenic pollutants?" *Journal of the National Cancer Institute* 89 (12): 888–89.

Dewailly, Eric, et al. 1994. "High organochlorine body burden in women with estrogen receptor-positive breast cancer." *Journal of the National Cancer Institute* 86 (3): 232–34.

Domchek, Susan M., et al. 2006. "Mortality after bilateral salpingo-oophorectomy in BRCA1 and BRCA2 mutation carriers: A prospective cohort study." *Lancet Oncology* 7 (3): 223–29.

Douglas, Mary, and Aaron Wildavsky. 1983. *Risk and Culture: An Essay on the Selection of Technological and Environmental Dangers.* Berkeley: University of California Press.

Dowie, Mark. 1995. *Losing Ground: American Environmentalism at the Close of the Twentieth Century.* Cambridge: MIT Press.

———. 1998. "What's wrong with the *New York Times*'s science reporting?" *Nation,* July 6, 13–19.

Duell, Eric J., et al. 2001. "Polymorphisms in the DNA repair gene *XRCC1* and breast cancer." *Cancer, Epidemiology, Biomarkers, and Prevention* 10 (3): 217–22.

Dumenil, Lynn. 1995. *The Modern Temper: American Culture and Society in the 1920s*. New York: Hill and Wang.

Duster, Troy. 1990. *Backdoor to Eugenics*. New York: Routledge.

Edlin, Gordon. 1987. "Inappropriate use of genetic terminology in medical research: A public health issue." *Perspectives in Biology and Medicine* 31 (Autumn): 47–56.

Egan, Kathleen M., et al. 1996. "Jewish religion and risk of breast cancer." *Lancet* 347 (9016): 1645–46.

Ehrenreich, Barbara, and Deirdre English. 1978. *For Her Own Good: 150 Years of Experts' Advice to Women*. New York: Doubleday.

Eisen, Andrea, and Barbara Weber. 1998. "Recent advances in breast cancer biology." *Current Opinion in Oncology* 10 (6): 486–91.

Eisen, Andrea, Timothy R. Rebbeck, William C. Wood, and Barbara L. Weber. 2000. "Prophylactic surgery in women with a hereditary predisposition to breast and ovarian cancer." *Journal of Clinical Oncology* 18 (9): 1980–95.

Eisenberg, Esther. 2006. "Comment." http://www.ovaryresearch.com/professional-responses.htm#nams (accessed October 30, 2012).

Elliott, Carl. 2001. "Pharma buys a conscience." *American Prospect* 12 (17): 16–20.

Elson, Jean. 2004. *Am I Still a Woman: Hysterectomy and Gender Identity*. Philadelphia: Temple University Press.

Engel, Leonard. 1962. "The race to create life." *Harper's*, October, 39–45.

Epstein, Samuel S. 1998. *The Politics of Cancer Revisited*. Fremont Center, NY: East Ridge.

Epstein, Steven. 1996. *Impure Science: AIDS, Activism, and the Politics of Knowledge*. Berkeley: University of California Press.

———. 2007. *Inclusion: The Politics of Difference in Medical Research*. Chicago: University of Chicago Press.

Evans, Nancy, ed. 2006. "State of the evidence: What is the connection between the environment and breast cancer?" San Fransisco: Breast Cancer Fund and Breast Cancer Action.

Evans, Sara M. 1989. *Born for Liberty: A History of Women in America*. New York: Free Press.

Falck, Frank, et al. 1992. "Pesticides and polychlorinated biphenyl residues in human breast lipids and their relation to breast cancer." *Archives of Environmental Health* 47 (March–April): 143–46.

Faludi, Susan. 1991. *Backlash: The Undeclared War against American Women*. New York: Crown.

Fausto-Sterling, Anne. 1985. *Myths of Gender: Biological Theories about Women and Men*. New York: Basic.

———. 2000a. "The five sexes, revisited." *Sciences* (July–August): 18–23.

———. 2000b. *Sexing the Body: Gender Politics and the Construction of Sexuality*. New York: Basic.

———. 2004. "Refashioning race: DNA and the politics of health care." *Differences* 15 (3): 1–37.

Feenberg, Andrew. 1991. *Critical Theory of Technology*. New York: Oxford University Press.

Feldman, Marcus W., Richard C. Lewontin, and Mary-Claire King. 2003. "A genetic melting-pot." *Nature* 424 (July 24): 374.

Fields, Barbara. 1990. "Slavery, race and ideology in the United States of America." *New Left Review* 1 (181): 95–118.

Fortun, Michael. 1995. "Mapping and making histories: The genomics project in the United States, 1980–1990." PhD diss., Harvard University.

Foucault, Michel. 1994 [1973]. *The Birth of the Clinic: An Archaeology of Medical Perception*. New York: Vintage.

Frank, Glenn. 1922. "An American looks at his world." *Century*, December, 317.

Freedman, Estelle B. 2002. *No Turning Back: The History of Feminism and the Future of Women*. New York: Ballantine.

Freeman, Harold P. 1998. "The meaning of race in science—Considerations for cancer research." *Cancer* 82 (1): 219–25.

Fregene, Alero, and Lisa A. Newman. 2005. "Breast cancer in sub-Saharan Africa: How does it relate to breast cancer in African-American women?" *Cancer* 103 (8): 1540–50.

Friend, Stephen H. 1996. "Breast cancer susceptibility testing: Realities in the post-genomic era." *Nature Genetics* 13 (1): 16–17.

Friend, Tim. 1994. "Inherited breast cancer gene located." *USA Today*, September 15.

Fry, Allison, R. Rush, C. Busby-Earle, A. Cull. 2001. "Deciding about prophylactic oophorectomy: What is important to women at increased risk of ovarian cancer?" *Preventive Medicine* 33 (6): 578–85.

Fujimura, Joan H., and Ramya Rajagopalan. 2011. "Different differences: The use of 'genetic ancestry' versus race in biomedical human genetic research." *Social Studies of Science* 41 (5): 5–30.

Gallion, Holly H., and Robert Park. 1995. "Developing intervention/prevention strategies for individuals at high risk of developing hereditary ovarian cancer." *Journal of the National Cancer Institute Monographs* (17): 103–6.

Gao, Qing, et al. 1997. "Recurrent germ-line *BRCA1* mutations in extended African American families with early-onset breast cancer." *American Journal of Human Genetics* 60 (5): 1233-36.

———. 2000a. "Prevalence of BRCA1 and BRCA2 mutations among clinic-based African American families with breast cancer." *Human Genetics* 107 (2): 186–91.

———. 2000b. "Protein truncating *BRCA1* and *BRCA2* mutations in African women with pre-menopausal breast cancer." *Human Genetics* 107: 192–94.

Garber, Judy Ellen, and Anne-Renee Hartman. 2004. "Prophylactic oophorectomy and hormone replacement therapy: Protection at what price?" *Journal of Clinical Oncology* 22 (6): 978–80.

Gearty, Robert. 1998. "Breast cancer and geography." *New York Daily News*, February 20.

Gehlert, Sarah, et al. 2008. "Targeting health disparities: A model linking upstream determinants to downstream interventions." *Health Affairs* 27 (2): 339–49.

Geiser, Ken. 1999. "Establishing a general duty of precaution in environmental protection policies in the United States." In *Protecting Public Health and the Environment: Implementing the Precautionary Principle*, edited by Carolyn Raffensperger and Joel Tickner, xxi–xxvi. Washington: Island.

"Geneticist warns on radiation rise: Takes sharp issue with views of Strauss and predicts human 'defective' increase." 1954. *New York Times*, September 12.

German Consortium for Hereditary Breast and Ovarian Cancer. 2002. "Comprehensive analysis of 989 patients with breast or ovarian cancer provides *BRCA1* and *BRCA2* mutation profiles and frequencies for the German population." *International Journal of Cancer* 97 (4): 472–80.

Gilbert, Walter. 1987. "Genome sequencing: Creating a new biology for the twenty-first century." *Issues in Science and Technology* 3 (3): 26–35.

Goldgar, David E., and Philip R. Reilly. 1995. "A common BRCA1 mutation in the Ashkenazim." *Nature Genetics* 11 (October): 113–14.

Goldstein, Marilyn. 1999. "Come and be counted: The story behind the Long Island Breast Cancer Study." *MAMM* (July–August):42–72.

Goodell, Rae. 1986. "How to kill a controversy: The case of recombinant DNA." In *Scientists and Journalists: Reporting Science as News*, edited by Sharon M. Friedman, Sharon Dunwoody, and Carol L. Rogers, 170–81. New York: Free Press.

Goodhue, Stoddard. 1913. "Do you choose your children?" *Cosmopolitan*, July 1913, 148–57.

Goodman, Annekathryn, and Karen Houck. 2001. "Anxiety and uncertainty in informed decision making." *Journal of Women's Health & Gender-Based Medicine* 10 (2): 93–94.

Goodman, Marc T., et al. 2003. "Incidence of ovarian cancer by race and ethnicity in the United States, 1992–1997." *Cancer* 97 (10 Supplement): 2676–85.

Gould, Stephen Jay. 1996. *The Mismeasure of Man*. New York: Norton.

Grady, Denise. 2006a. "Imperfect, imprecise, but useful: your race." *New York Times* July 4.

———. 2006b. "Racial component is found in lethal breast cancer." *New York Times*, June 7.

Graham, Loren. 1993. *Science in Russia and the Soviet Union: A Short History*. Cambridge: Cambridge University Press.

Grann, Victor R., et al. 2002. "Effect of prevention strategies on survival and quality-adjusted survival of women with BRCA 1/2 mutations: An updated decision analysis." *Journal of Clinical Oncology* 20 (10): 2520–29.

Gray, Janet, Janet Nudelman, and Connie Engel. 2010. *State of the Evidence: The Connection between Breast Cancer and the Environment*. http://www.

breastcancerfund.org/media/publications/state-of-the-evidence/ (accessed November 9, 2012).

Guillemette, Chantal, Robert C. Millikan, Beth Newman, and David E. Housman. 2000. "Genetic polymorphisms in uridine diphospho-glucuronosyltransferase 1A1 and association with breast cancer among African Americans." *Cancer Research* 60 (4): 950–56.

Haber, Daniel. 2002. "Prophylactic oophorectomy to reduce the risk of ovarian and breast cancer in carriers of BRCA mutations." *New England Journal of Medicine* 346 (21): 1660–62.

Hacking, Ian. 1983. *Representing and Intervening: Introductory Topics in the Philosophy of Science.* Cambridge: Cambridge University Press.

Haffty, B. G., et al. 2006. "Racial differences in the incidence of BRCA1 and BRCA2 mutations in a cohort of early onset breast cancer patients: African American compared to white women." *Journal of Medical Genetics* 43 (2): 133–37.

———. 2009. "Breast cancer in young women (YBC): Prevalence of BRCA1/2 mutations and risk of secondary malignancies across diverse racial groups." *Annals of Oncology* 20 (10): 1653–59.

Hall, Jason J., and Don J. Hall. 2006. "The forgotten hysterectomy: The first successful abdominal hysterectomy and bilateral salpingo-oophorectomy in the United States." *Obstetrics and Gynecology* 107 (2): 541–43.

Hall, Jeff M., et al. 1990. "Linkage of early-onset familial breast cancer to chromosome 17q21." *Science* 250 (December 21): 1684–89.

Hall, Michael J., and Olufunmilayo I. Olopade. 2006. "Disparities in genetic testing: Thinking outside the BRCA box." *Journal of Clinical Oncology* 24 (14): 2197–203.

Hall, Michael J., et al. 2009. "BRCA1 and BRCA2 mutations in women of different ethnicities undergoing testing for hereditary breast-ovarian cancer." *Cancer* 115 (May 15): 2222–33.

Haller, Mark. 1984. *Eugenics: Hereditarian Attitudes in American Thought.* New Brunswick: Rutgers University Press.

Hallowell, Nina. 1998. "You don't *want* to lose your ovaries because you think 'I might become a man':. Women's perceptions of prophylactic surgery as a cancer risk management option." *Psycho-Oncology* 7:263–75.

———et al. 2001. "Surveillance or surgery? A description of the factors that influence high risk premenopausal women's decisions about prophylactic oophorectomy." *Journal of Medical Genetics* 38 (10): 683–91.

Happe, Kelly. 2000. "Race betterment at the turn of the century, or, why it's OK to marry your cousin." In *Turning the Century: Essays in Media and Cultural Studies*, edited by Carol A. Stabile, 166–86.. Boulder, CO: Westview.

———. 2006. "Heredity, gender, and the discourse of ovarian cancer." *New Genetics and Society* 25 (2): 171–96.

Haraway, Donna. 1989. "The biopolitics of postmodern bodies: Determinations of self in immune system discourse." *Differences* 1 (1): 3–43.

Harding, Sandra. 1991. *Whose Science? Whose Knowledge? Thinking from Women's Lives*. Ithaca: Cornell University Press.

———. 2006. *Science and Social Inequality: Feminist and Postcolonial Issues*. Urbana: University of Illinois Press.

Hartmann, Heidi. 1997. "The unhappy marriage of Marxism and feminism: Towards a more progressive union." In *The Second Wave: A Reader in Feminist Theory*, edited by Linda Nicholson, 97–122. New York: Routledge.

Hartmann, Lynn C., et. al. 1999. "Efficacy of bilateral prophylactic mastectomy in women with a family history of breast cancer." *New England Journal of Medicine* 340 (2): 77–138.

Hayles, N. Katherine. 1993. "The materiality of informatics." *Configurations* 1 (1): 147–70.

Heffner, Linda J. 2004. "Advanced maternal age: How old is too old?" *New England Journal of Medicine* 351 (19): 1927–29.

Hickey, M., M. Ambekar, and I. Hammond, I. 2010. "Should the ovaries be removed or retained at the time of hysterectomy for benign disease?" *Human Reproduction Update* 16 (2): 131–41.

Higham, John. 1988. *Strangers in the Land: Patterns of American Nativism, 1860–1925*. New Brunswick: Rutgers University Press.

Holmes, Robert H. 1955. "Report from Hiroshima: Latest about after-effects of A-bomb." *U.S. News and World Report*, May 13, 60–68.

Holt, Jeffrey T., et al. 1996. "Growth retardation and tumour inhibition by BRCA1." *Nature Genetics* 12 (March): 298–302.

Hotz, Robert Lee. 1995. "Gene defect may provide cancer test." *Los Angeles Times*, September 29.

Høyer, Annette, et al. 1998. "Organochlorine exposure and risk of breast cancer." *Lancet* 352 (December 5): 1816–20.

Hu, Jennifer J., et al. 2002. "Genetic regulation of ionizing radiation sensitivity and breast cancer risk." *Environmental and Molecular Mutagenesis* 39 (2–3): 208–15.

Hubbard, Ruth. 2003. "Science, power, gender: How DNA became the book of life." *Signs* 28 (3): 791–99.

——— and Mary S. Henifin. 1985. "Genetic screening of prospective parents and of workers: Some scientific and social issues." *International Journal of Health Services* 15 (2):231–51.

Huggins, Charles, and Thomas L. Y. Dao. 1953. "Adrenalectomy and oophorectomy in treatment of advanced carcinoma of the breast." *Journal of the American Medical Association* 151 (16): 1388–94.

Huisman, Wouter M. 1997. "Prophylactic oophorectomy on post-menopausal women." *Human Reproduction* 12 (2): 204–5.

Hunter, David J., and Karl T. Kelsey. 1993. "Pesticide residues and breast cancer: The harvest of a silent spring?" *Journal of the National Cancer Institute* 85 (8): 598–99.

Hunter, David J., et al. 1997. "Plasma organochlorine levels and the risk of breast cancer." *New England Journal of Medicine* 337 (18):1253–58.

Huo, Dezheng, and Olufunmilayo I. Olopade. 2007. "Genetic testing in diverse populations: Are researchers doing enough to get out the correct message?" *Journal of the American Medical Association* 298 (24): 2910–11.

Hurley, Karen E., et al. 2001. "Anxiety/uncertainty reduction as a motivation for interest in prophylactic oophorectomy in women with a family history of ovarian cancer." *Journal of Women's Health & Gender-Based Medicine* 10 (2): 189–99.

Husted, Amanda. 1994. "No link found between breast cancer and exposure to pesticides, study says." *Atlanta Journal-Constitution*, April 20.

Institute of Medicine. 2001. *Health and Behavior: The Interplay of Biological, Behavioral, and Societal Influences.* Washington: National Academies Press.

International HapMap Consortium. 2005. "A haplotype map of the human genome." *Nature* 437 (October 27): 1299–1320.

International HapMap Project. 2012. "How will the HapMap benefit human health?" http://hapmap.ncbi.nlm.nih.gov/healthbenefit.html.en (last accessed on November 2, 2012).

"Is race real? A Web forum organized by the Social Science Research Council." 2005. http://raceandgenomics.ssrc.org (accessed October 28, 2012).

Jackson, Fatimah. 1998. "Scientific limitations and ethical ramifications of a non-representative human genome project: African American response." *Science and Engineering Ethics* 4 (2): 155–70.

Jackson, Richard. 2000. "Unburdening ourselves." *Silent Spring Review*, Fall, 6–7. On file with author.

Jameel, J. K. A., V. S. R. Rao, L. Cawkwell, P. J. Drew. 2004. "Radioresistance in carcinoma of the breast." *Breast* 13 (6):452–60.

Jasanoff, Sheila. 2004. "The idiom of co-production." In *States of Knowledge: The Co-Production of Science and Social Order*, edited by Sheila Jasanoff, 1–12. London: Routledge.

Jenks, Susan. 1994. "Researchers to comb Long Island for potential cancer factors." *Journal of the National Cancer Institute* 86 (2): 88–89.

Jensen, Roy A., et al. 1996. "BRCA1 is secreted and exhibits properties of a granin." *Nature Genetics* 12 (March): 303–8.

John, Esther M., et al. 2007. "Prevalence of pathogenic BRCA1 mutation carriers in 5 US racial/ethnic groups." *Journal of the American Medical Association* 298 (24): 2869–76.

Jones, Lovell A., and Janice A. Chilton. 2002. "Impact of breast cancer on African-American women: Priority areas for research in the next decade." *American Journal of Public Health* 92 (4): 539–42.

Jordan, Andrew, and Timothy O'Riordan. 1999. "The precautionary principle in contemporary environmental policy and politics." In *Protecting Public Health and the Environment: Implementing the Precautionary Principle*, edited by Carolyn Raffensperger and Joel Tickner, 15–35.. Washington: Island.

Juengst, Eric T. 1996. "Self-critical federal science? The ethics experiment within the U.S. Human Genome Project." In *Scientific Innovation, Philosophy, and Public Policy*, edited by Ellen Frankel Paul, Fred D. Miller, and Jeffrey Paul, 63–95. New York: Cambridge University Press.

Kaempffert, Waldemar. 1956. "Science in review: National Academy report on radiation poses problems faced by humanity." *New York Times*, June 17.

Kahn, Johnathan. 2004. "How a drug becomes 'ethnic': Law, commerce, and the production of racial categories in medicine." *Yale Journal of Health Policy, Law, and Ethics* 4 (1): 1–46.

Kanaan, Yasmine, et al. 2003. "Inherited BRCA2 mutations in Aftrican Americans with breast and/or ovarian cancer: A study of familial and early onset cases." *Human Genetics* 113 (5): 452–60.

Karlan, Beth Y. 2004. "Defining cancer risks for BRCA germline mutation carriers: Implications for surgical prophylaxis." *Gynecologic Oncology* 92 (2): 519–20.

Kauff, Noah D., and Richard Barakat. 2004. "Surgical risk-reduction in carriers of BRCA mutations: Where do we go from here?" *Gynecologic Oncology* 93 (2): 277–79.

———. 2007. "Risk-reducing salpingo-oophorectomy in patients with germline mutations in BRCA1 or BRCA2." *Journal of Clinical Oncology* 25 (20): 2921–27.

Kauff, Noah D., et al. 2002. "Risk-reducing salpingo-oophorectomy in women with a BRCA1 or BRCA2 mutation." *New England Journal of Medicine* 346 (21):1609–15.

Kaufman, Jay S., Richard S. Cooper, and Daniel L. McGee. 1997. "Socioeconomic status and health in blacks and whites: The problem of residual confounding and the resiliency of race." *Epidemiology* 8 (6): 621–28.

Kay, Lily. 1993. *The Molecular Vision of Life: Caltech, The Rockefeller Foundation, and the Rise of the New Biology*. New York: Oxford University Press.

———. 2000. *Who Wrote the Book of Life? A History of the Genetic Code*. Stanford: Stanford University Press.

Keller, Evelyn Fox. 1992a. "Nature, nurture, and the Human Genome Project." In *The Code of Codes: Scientific and Social Issues in the Human Genome Project*, edited by Daniel Kevles and Leroy Hood, 281–99. Cambridge: Harvard University Press.

———. 1992b. *Secrets of Life, Secrets of Death: Essays on Language, Gender and Science*. New York: Routledge.

———. 2000. *The Century of the Gene*. Cambridge: Harvard University Press.

Keoun, Brad. 1997. "Ashkenazim not alone: Other ethnic groups have breast cancer gene mutations, too." *Journal of the National Cancer Institute* 89 (1): 8–9.

Kevles, Daniel. 1985. *In the Name of Eugenics: Genetics and the Uses of Human Heredity*. Berkeley: University of California Press.

———. 1992. "Out of eugenics: The historical politics of the human genome." In *The Code of Codes: Scientific and Social Issues in the Human Genome Project*, edited by Daniel Kevles and Leroy Hood, 3–36. Cambridge: Harvard University Press.

———. 1997. "Big science and big politics in the United States: Reflections on the death of the SSC and the life of the Human Genome Project." *Historical Studies in the Physical and Biological Sciences* 27 (2): 269–97.

Khanna, Kum. 2000. "Cancer risk and the ATM gene: A continuing debate." *Journal of the National Cancer Institute* 92 (10): 795–802.

Khoury, Muin J. 2001. "Message from Muin J. Khoury M.D., Ph.D., cirector, National Office of Public Health Genomics." CDC Office of Surveillance, Epidemiology, and Laboratory Services, Public Health Genomics. www.cdc.gov/genomics/about/welcome.htm (accessed October 28, 2012

———. 2003. "Genomics research in the 21st century: From the test tube to population health." www.hhs.gov/asl/testify/t030522b.html (accessed October 28, 2012).

Kim, Lillian Lee. 1998. "Breast cancer risk factors appear higher in lesbians." *Atlanta Journal-Constitution*, September 30.

King, Mary-Claire, Joan H. Marks, Jessica B. Mandell. 2003. "Breast and ovarian cancer risks due to inherited mutations in BRCA1 and BRCA2." *Science* 302 (October 24): 643–46.

King, Mary-Claire, et al. 2001. "Tamoxifen and breast cancer incidence among women with inherited mutations in BRCA1 and BRCA2." *Journal of the American Medical Association* 286 (18): 2251–56.

Kinney, Anita Yeomans, et al. 2001. "Knowledge, attitudes, and interest in breast-ovarian cancer gene testing: A survey of large African-American kindred with a BRCA1 mutation." *Preventive Medicine* 33 (6): 543–51.

———. 2005. "The impact of receiving genetic test results on general and cancer-specific psychologic distress among members of an African-American kindred with a *BRCA1* mutation." *Cancer* 104 (11): 2508–16.

Klawiter, Maren. 2008. *The Biopolitics of Breast Cancer: Changing Cultures of Disease and Activism*. Minneapolis: University of Minnesota Press.

Kline, Wendy. 2001. *Building a Better Race: Gender, Sexuality, and Eugenics from the Turn of the Century to the Baby Boom*. Berkeley: University of California Press.

Knudson, Alfred G., Jr. 1971. "Mutation and cancer: statistical study of retinoblastoma." *Proceedings of the National Academy of Sciences in the United States of America* 68 (4): 820–23.

Koenig, Barbara A., et al. 1998. "Genetic testing for BRCA1 and BRCA2: Recommendations of the Stanford Program in Genomics, Ethics, and Society." *Journal of Women's Health* 7 (5): 531–45.

Kolata, Gina. 2002. "Breast cancer on Long Island: No epidemic despite clamor for action." *New York Times*, August 29.

———. 2009. "In long drive to cure cancer, advances have been elusive." *New York Times*. April 24.

Koshland, Daniel E. 1989. "The sequence and consequences of the human genome."
   Science 246 (October 13): 189.

Krieger, Nancy. 1989. "Exposure, susceptibility, and breast cancer risk: A hypothesis regarding exogenous carcinogens, breast tissue development, and social
   gradients, including black/white differences, in breast cancer incidence." *Breast
   Cancer Research and Treatment* 13 (3): 205–23.

—— et al. 1994a. "Breast cancer and serum organochlorines: A prospective study
   among white, black, and Asian women." *Journal of the National Cancer Institute*
   86 (8): 589–99.

——. 1994b. "Response." *Journal of the National Cancer Institute* 86 (14): 1094–95.

——. 1999. "Embodying inequality: a review of concepts, measures, and methods
   for studying health consequences of discrimination." *International Journal of
   Health Services* 29 (2): 295–352.

——. 2001. "Theories for social epidemiology in the 21st century: An ecosocial
   perspective." *International Journal of Epidemiology* 30 (4): 668–77.

——. 2002. "Is breast cancer a disease of affluence, poverty, or both? The case of
   African-American women." *American Journal of Public Health* 92 (4): 611–13.

——. 2005. "Embodiment: A conceptual glossary for epidemiology." *Journal for
   Epidemiology and Community Health* 59 (5): 350–55.

——. 2011. *Epidemiology and the People's Health: Theory and Context.* Oxford:
   Oxford University Press.

Krimsky, Sheldon. 1982. *The Social History of the Recombinant DNA Controversy.*
   Cambridge: MIT Press.

——. 2000. *Hormonal Chaos: The Scientific and Social Origins of the Environmental Endocrine Hypothesis.* Baltimore: Johns Hopkins University Press.

Kuehn, Robert R. 1996. "The environmental justice implications of quantitative risk
   assessment." *University of Illinois Law Review,* no. 1, 103–72.

Kuhn, Thomas S. 1970 [1962]. *The Structure of Scientific Revolutions.* Chicago: University of Chicago Press.

Kutz, Fredrick., A. R. Yobs, and S. C. Strassman. 1977. "Racial stratification of
   organiclorine insecticide residue in human adipose tissue." *Journal of Occupational Medicine* 9 (9): 619–22.

Lan, Q. et al., 2004. Hematotoxicity in workers exposed to low levels of benzene.
   Science 306 (December): 1774-1776.

Lan, Qing et al., 2009. Large-scale evaluation of candidate genes identifies associations between DNA repair and genomic maintenance and development of
   benzene hematotoxicity. *Carcinogenesis* 30, no. 1: 50-58.

Laurence, William. 1962. "Importance of the discovery of the structure of DNA is
   examined." *New York Times,* October 21.

Lavin, Martin, et al. 1994. "Identification of a potentially radiosensitive subgroup
   among patients with breast cancer." *Journal of the National Cancer Institute* 86
   (21): 1627–34.

Lee, Thomas F. 1991. *The Human Genome Project: Cracking the Genetic Code of Life.* New York: Plenum.

Leroi, Armand Marie. 2005. "A family tree in every gene." *New York Times,* March 14.

Leviero, Anthony. 1956. "Scientists term radiation a peril to future of man; even small dose can prove harmful to descendants of victim, report states." *New York Times,* June 13.

Levin, David, ed. 1993. *Modernity and the Hegemony of Vision.* Berkeley: University of California Press.

Levine, Douglas, et al. 2003. "Fallopian tube and primary peritoneal carcinoma associated with BRCA mutations." *Journal of Clinical Oncology* 21 (22): 4222–27.

Levins, Richard. 2000. "Is capitalism a disease? The crisis in U.S. public health." *Monthly Review* 52 (4): 8–33.

——— and Richard Lewontin. 1985. *The Dialectical Biologist.* Cambridge: Harvard University Press.

Lewine, Howard. 2005. "News review from Harvard Medical School: Keep ovaries, researchers say." *OvaryResearch.com.* http://www.ovaryresearch.com/profes-sional-responses.htm (accessed October 28, 2012).

Lewontin, Richard C. 1972. "The apportionment of human diversity." *Evolutionary Biology* 6:381–98.

———. 1991. *Biology as Ideology: The Doctrine of DNA.* New York: HarperPerennial.

———, and Richard Levins. 2007. *Biology under the Influence: Dialectical Essay on Ecology, Agriculture, and Health.* New York: Monthly Review Press.

———, Steven Rose, and Leon J. Kamin. 1985. *Not in Our Genes: Biology, Ideology, and Human Nature.* New York: Pantheon.

Ley, Barbara. 2009. *From Pink to Green: Disease Prevention and the Environmental Breast Cancer Movement.* New Brunswick: Rutgers University Press.

Liede, Alexander, et al. 2002. "Cancer incidence in a population of Jewish women at risk of ovarian cancer." *Journal of Clinical Oncology* 20 (6): 1570–77.

Lindsey, Heather. 2006. "Study: Prophylactic oophorectomy may increase mortal-ity in some patients at average risk of ovarian cancer." *Oncology Times* 28 (24): 21–24.

Lipkus, Isaac M., Deborah Iden, Jennifer Terrenoire, and John R. Feaganes. 1999. "Relationships among breast cancer concern, risk perceptions, and interest in genetic testing for breast cancer susceptibility among African-American women with and without a family history of breast cancer." *Cancer, Epidemology, Bio-markers, and Prevention* 8 (June): 533–39.

Lippman, Abby. 1991. "Prenatal genetic testing and screening: Constructing needs and reinforcing inequities." *American Journal of Law and Medicine* 17 (1–2): 15–49.

Locke, Paul, and D. Bruce Myers Jr. 2010. "Food for thought . . . a replacement-first approach to toxicity testing is necessary to successfully reauthorize TSCA." *Alternatives to Animal Experimentation* 28 (4): 266–72.

Long, Jeffrey C., and Rick A. Kittles. 2003. "Human genetic diversity and the non-existence of biological races." *Human Biology* 75 (4): 449–71.

Longino, Helen. 1990. *Science and Social Knowledge: Values and Objectivity in Scientific Inquiry*. Princeton: Princeton University Press.

Love, Richard R., and John Philips. 2002. "Oophorectomy for breast cancer: history revisited." *Journal of the National Cancer Institute* 94 (19): 1433–34.

Lowe, Donald M. 1995. *The Body in Late-Capitalist USA*. Durham: Duke University Press.

Ludmerer, Kenneth M. 1972. *Genetics and American Society: A Historical Appraisal*. Baltimore: Johns Hopkins University Press.

Mamon, H. J., et al. 2003. "Differing effects of breast cancer 1, early onset (BRCA1) and ataxia-telangiectasia mutated (ATM) mutations on cellular responses to ionizing radiation." *International Journal of Radiation* 79 (10): 817–29.

Markowitz, Gerald, and David Rosner. 2002. *Deceit and Denial: The Deadly Politics of Industrial Pollution*. Berkeley: University of California Press.

Marks, Jonathan, 2006. "The realities of races." In "Is race 'real'? A web forum organized by the Social Science Research Council." June 7. http://raceandgenomics. ssrc.org (accessed October 29, 2012).

Marquis, Sandra T., et al. 1995. "The developmental pattern of BRCA1 expression implies a role in differentiation of the breast and other tissues." *Nature Genetics* 11 (September): 17–26.

Marsh, Margaret, and Wanda Ronner. 2008. *The Fertility Doctor: John Rock and the Reproductive Revolution*. Baltimore: John Hopkins University Press.

Martin, Anne-Marie, and Barbara Weber. 2000. "Genetic and hormonal risk factors in breast cancer." *Journal of the National Cancer Institute* 92 (14): 1126–35.

Martin, Emily. 1987. *The Woman in the Body: A Cultural Analysis of Reproduction*. Boston: Beacon.

———. 1990. "The end of the body?" *American Ethnologist* 19 (1): 121–40.

———. 1994. *Flexible Bodies: The Role of Immunity in American Culture from the Days of Polio to the Age of AIDS*. Boston: Beacon.

Martin, Joyce A. 2012. "Births: final data for 2010." National Vital Statistics Reports. National Center for Health Statistics. 61(1). August. http://www.cdc.gov/nchs/ data/nvsr/nvsr61/nvsr61_01.pdf (last accessed on November 9, 2012).

Martinot, Steven. 2007. "Motherhood and the invention of race." *Hypatia* 22 (2007): 81.

Maugh, Thomas H., II. 1994a."Breast cancer study clears DDT, PCBs: New research by Kaiser contradicts earlier findings; But the report in no way suggests that these chemicals are safe, an epidemiologist warns." *Los Angeles Times*, April 20.

———. 1994b. "Discovery of breast cancer gene called major advance: Researchers locate defect that causes half of inherited cases; Early detection may save many lives." *Los Angeles Times*, September 15.

Mayo Clinic. 2006. "Preventive ovary removal linked to early death in younger women, Mayo Clinic discovers." Press release, September 13. On file with author.

Mazumdar, Pauline. 1992. *Eugenics, Human Genetics, and Human Failings: The Eugenics Society and Its Critics in Britain.* New York: Routledge.

McClintock, Martha K., et al. 2005. "Mammary cancer and social interactions: Identifying multiple environments that regulate gene expression throughout the life span." *Journals of Gerontology: Series B* 60 (Spec. 1): 32–41.

McCormick, Sabrina. 2010. *No Family History: The Environmental Links to Breast Cancer.* Roman and Littlefield.

McGregor, Douglas H., et al. 1977. "Breast cancer incidence among atomic bomb survivors, Hiroshima and Nagasaki." *Journal of the National Cancer Institute* 59 (3): 799–811.

McMahon, Connette P. 2000. "Ephraim McDowell, Jane Todd Crawford, and the origins of oophorectomy." *North Carolina Medical Journal* 61 (1): 401–2.

Medina, Daniel, et al. 2002. "Environmental carcinogens and *p53* tumor-suppressor gene interactions in a transgenic mouse model for mammary carcinogenesis." *Environmental and Molecular Mutagenesis* 39 (2–3): 178–83.

Mefford, Heather C., et al. 1999. "Evidence for a BRCA1 founder mutation in families of West African ancestry." *American Journal of Human Genetics* 65 (2): 575–78.

Metzl, Jonathan M., and Anna Kirkland. 2010. *Against Health: How Health Became the New Morality.* New York: New York University Press.

Miki, Yoshio, et al. 1994. "A strong candidate for the breast and ovarian cancer susceptibility gene BRCA1." *Science* 266 (October 7): 66–71.

Miller, Elizabeth Kiggen. 1998. "Proposals awaited for L.I. breast cancer study." *New York Times*, August 9.

Miller, Suzanne M., et al. 2005. "Enhanced counseling for women undergoing BRCA 1/2 testing: Impact on subsequent decision making about risk reduction behaviors." *Health Education & Behavior* 32 (5): 654–67.

Millikan, Robert, et al. 2000. "Dichlorodiphenyldichloroethene, polychlorinated biphenyls, and breast cancer among African-American and white women in North Carolina." *Cancer Epidemiology, Biomarkers, and Prevention* 9 (11): 1233–40.

Miranda, Marie Lynn, and Dana C. Dolinoy. 2005. "Using GIS-based approaches to support research on neurotoxicants and other children's environmental health threats." *Neurotoxicology* 26 (2): 223–28.

Mitchell, Gordon, and Kelly Happe. 2001 "Informed consent after the Human Genome Project." *Rhetoric and Public Affairs* 4 (3): 375–406.

Montagu, Ashley, 1964. *The Concept of Race.* New York: Free Press.

Morantz-Sanchez, Regina. 1999. *Conduct Unbecoming a Woman: Medicine on Trial in Turn-of-the-Century Brooklyn.* Oxford: Oxford University Press.

Morello-Frosch, Rachel. 1997. "The politics of reproductive hazards in the workplace: Class, gender, and the history of occupational lead exposure." *International Journal of Health Services* 27 (3): 501–21.

Morgan, Lynn M., and Meredith W. Michaels, eds. 1999. *Fetal Subjects, Feminist Positions.* Philadelphia: University of Pennsylvania Press.

Moscucci, Ornella. 1990. *The Science of Woman: Gynaecology and Gender in England, 1800–1929*. Cambridge: Cambridge University Press.

——— and Aileen Clarke. 2007. "Prophylactic oophorectomy: A historical perspective." *Journal of Epidemiological Community Health* 61 (3): 182–84.

Mosher, William D., and Jo Jones. 2010. "Use of contraception in the United States: 1982–2008." *Vital Health Statistics* 23 (29).

Muller, Hermann J. 1955. "What will radioactivity do to our children? An interview with Dr. H. J. Muller, Nobel Prize winner in genetics." *U.S. News and World Report*, May 13, 72–78.

Mullineaux, Lisa.G., et al. 2003. "Identification of germline 185delAG BRCA1 mutations in non-Jewish Americans of Spanish ancestry from the San Luis Valley, Colorado." *Cancer* 98 (3): 597–602.

Murray, Stuart J. 2006. "Thanatopolitics: On the use of death for mobilizing political life." *Polygraph* 18:191–215

———. 2007. "Care and the self: Biotechnology, reproduction, and the good life." *Philosophy, Ethics, and Humanities in Medicine* 2 (6): 1–15.

Mussalo-Rauhamaa, H., et al. 1990. "Occurrence of beta-hexachlorocyclohexane in breast cancer patients." *Cancer* 66 (10): 2124–28.

Nanda, Rita, et al. 2005. "Genetic testing in an ethnically diverse cohort of high-risk women." *Journal of the American Medical Association* 294 (15): 1925–33.

Narod, S. A., et al. 1991. "Familial breast-ovarian cancer locus on chromosome 17q12-q23." *Lancet* 338 (8759): 82–83.

National Breast Cancer Coalition Fund. 2000. "Project Lead (Leadership, Education, and Advocacy Development): An innovative science training program for breast cancer activists." On file with author.

National Cancer Institute. 2011a. "Breast cancer." http://www.cancer.gov/cancertopics/types/breast (accessed July 2011). On file with author.

———. 2011b. "Ovarian Cancer." http://www.cancer.gov/cancertopics/types/ovarian (accessed July 2011). On file with author.

———. 2012. PDQ® genetics of breast and ovarian cancer." Bethesda, MD: National Cancer Institute. http://cancer.gov/cancertopics/pdq/genetics/breast-and-ovarian/HealthProfessional (accessed November 9, 2012).

——— and National Institute of Environmental Health Sciences. 2000. "Long Island Breast Cancer Study Project: Interim report." June. http://epi.grants.cancer.gov/LIBCSP/InterimReport/index.html?view=plain (accessed October 28, 2012).

National Center for Environmental Health. 2001. "National report on human exposure to environmental chemicals." Atlanta, GA: Centers for Disease Control and Prevention. March 21. On file with author.

National Human Genome Research Institute 2011. "Genes, enviroment, and health initiative." October. http://www.genome.gov/19518663 (accessed March 3, 2012).

National Institute for Environmental Health Sciences. 2000. "Environment Health Institute's Centers to breed mice with human-like gene variants that modify their responses to environmental factors and the repair of damaged DNA." January 3. On file with author.

———. 2008. "Environmental Genome Project: Program description."November 14. On file with author.

———. N.d. "Environmental Genome Project: Genotyping background." http://egp. gs.washington.edu/genotyping_background.html (accessed November 9, 2012).

National Research Council. 2012. *Breast Cancer and the Environment: A Life Course Approach*. Washington: National Academies Press. http://www.iom.edu/ Reports/2011/Breast-Cancer-and-the-Environment-A-Life-Course-Approach. aspx (accessed November 9, 2012).

Nelkin, Dorothy, and M. Susan Lindee. 1995. *The DNA Mystique: The Gene as Cultural Icon*. New York: W. H. Freeman.

Nelson, Nancy J. 1998. "Ashkenazi community is not unwilling to participate in genetic research." *Journal of the National Cancer Institute* 90 (12): 884–85.

Newill, Vaun A. 1961. "Distribution of cancer mortality among ethnic subgroups of the white population of New York City, 1953–1958." *Journal of the National Cancer Institute* 26 (February):405–17.

Newman, Beth, et al. 1998. "Frequency of breast cancer attributable to BRCA1 in a population-based series of American women." *Journal of the American Medical Association* 279 (12): 915–21.

Newman, Laura. 2001. "Prophylactic oophorectomy in the genome age: Balancing new data against uncertainties." *Journal of the National Cancer Institute* 93 (3): 173–75.

Nicholson, Linda. 1997. "Feminism and Marx: Integrating kinship with the economic." In *The Second Wave: A Reader in Feminist Theory*, edited by Linda Nicholson, 131–46. New York: Routledge.

Office of Technology Assessment. 1988. "Mapping our genes—genome projects: How big? How Fast?" April. http://www.ornl.gov/sci/techresources/Human_ Genome/publicat/OTAreport.pdf(accessed October 29, 2012).

Ohmann, Richard. 1996. *Selling Culture: Magazines, Markets, and Class at the Turn of the Century*. New York: Verso.

Olby, Robert C. 1974. *Path to the Double Helix: The Discovery of DNA*. Seattle: University of Washington Press.

———. 1990. "The molecular revolution in biology." In *Companion to the History of Modern Science*, edited by Robert C. Olby G. N. Cantor, J. R. R. Christie, and M. J. S. Hodge, 503–19. London: Routledge.

Olive, David L. 2005. "Dogma, skepsis, and the analytic method: the role of prophylactic oophorectomy at the time of hysterectomy." *Obstetrics and Gynecology* 106 (2): 214–15.

Olopade, Olufunmilayo I. 2004. "Genetics in clincial cancer care: A promise unfulfilled among minority populations." *Cancer Epidemiology, Biomarkers, and Prevention* 13 (11): 1683–86.

——— and Grazia Artioli. 2004. "Efficacy of risk-reducing salpingo-oophorectomy in women with BRCA-1 and BRCA-2 mutations." *Breast Journal* 10 (Supplement 1): S5–9.

Olopade, Olufunmilayo I., et al. 2003. "Breast cancer genetics in African Americans." *Cancer* 97(Supplement 1): 236–45.

Omi, Michael, and Howard Winant. 1994. *Racial Formation in the United States.* London: Routledge.

Ordover, Nancy. 2003. *American Eugenics: Race, Queer Anatomy, and the Science of Nationalism.* Minneapolis: University of Minnesota Press.

Osborn, Peter, and Lynne Segel. 1994. "Gender as performance: An interview with Judith Butler." *Radical Philosophy* 67: 32–39.

Osmundsen, John A. 1962. "Biologists hopeful of solving secrets of heredity this year." *New York Times*, February 2.

Oyama, Susan, Paul E. Griffiths, and Russell D. Gray. 2001. *Cycles of Contingency: Developmental Systems and Evolution.* Cambridge: MIT Press.

Painter, Kim. 1994. "Implications of breast cancer gene." *USA Today*, September 15.

Pal, Tuya, Jenny Permuth-Wey, Tricia Holtje, and Rebecca Sutphen. 2004. "BRCA1 and BRCA2 mutations in a study of African American breast cancer patients." *Cancer Epidemiology, Biomarkers and Prevention* 13 (11): 1794–99.

Palmer, Julie R., et al. 2003. "Dual effect of parity on breast cancer risk in African-American women." *Journal of the National Cancer Institute* 95 (6): 478–83.

Panguluri, Ramesh C. K.,et al. 1999. "BRCA1 mutations in African Americans." *Human Genetics* 105 (1–2): 28–31.

Parham, Groesbeck, et al. 1997. "The National Cancer Data Base report on malignant epithelial ovarian carcinoma in African-American women." *Cancer* 80 (4): 816–26.

Parker, William H., et al. 2005a. "In Reply." *Obstetrics and Gynecology* 106 (5, Part 1): 1107.

———. 2005b. "Ovarian conservation at the time of hysterectomy for benign disease." *Obsetetrics and Gynecology* 106 (2): 219–25.

———2006. "Response to commentaries on retention of the ovaries and long-term survival after hysterectomy." *Climacteric* 9:390–400.

Parthasarathy, Shobita. 2007. *Building Genetic Medicine: Breast Cancer, Technology, and the Comparative Politics of Health Care.* Cambridge: MIT Press.

Patterson, James T. 1987. *The Dread Disease: Cancer and Modern American Culture.* Cambridge: Harvard University Press.

Patton, Sharyle, and Davis Baltz. 2005. "Taking it all in: Documenting chemical pollution in Californians through biomonitoring." August. http://www.commonweal.org/programs/brc/Taking_It_All_In.html (accessed November 9, 2012).

Paul, Diane B. 1995. *Controlling Human Heredity: 1865 to the Present.* Atlantic Highlands, NJ: Humanities Press International.

Petersen, Alan, and Deborah Lupton. 1996. *The New Public Health: Health and Self in the Age of Risk*. London: Sage.

Pezzullo, Phaedra. 2007. *Toxic Tourism: Rhetorics of Pollution, Travel, and Environmental Justice*. Tuscaloosa: University of Alabama Press.

Piver, M. Steven, and C. Wong. 1997. "Prophylactic oophorectomy: A century-long dilemma." *Human Reproduction* 12 (2): 205–6.

Piver, M. Steven, et al. 1993a. "Familial ovarian cancer: a report of 658 families from the Gilda Radner Familial Ovarian Cancer Registry 1981–1991." *Cancer* 71 (2 Supplement): 582–88.

———. 1993b. "Primary peritoneal carcinoma after prophylactic oophorectomy in women with a family history of ovarian cancer: A report of the Gilda Radner Familial Ovarian Cancer Registry." *Cancer* 71 (9): 2751–55.

Platt, Rutherford. 1962. "DNA: The mysterious basis of life." *Reader's Digest*, October, 141–48.

Proctor, Robert. 1995. *Cancer Wars: How Politics Shapes What We Know and Don't Know about Cancer*. New York: Basic.

Queller, Jessica. 2008. "To cut my breasts off, or not to cut my breasts off . . ." Interview with Corrie Pikul. *Salon.com*. http://www.salon.com/2008/04/02/jessica_queller/singleton/ (accessed October 29, 2012).

Rabin, Roni. 1995. "Researchers to study homes of women with breast cancer." *Houston Chronicle*, October 19.

Rabinow, Paul. 1992. "Artificiality and enlightenment: From sociobiology to biosociality." In *Incorporations*, edited by Jonathon Crary and Sanford Kwinter, 234–52. New York: Zone.

——— and Nikolas Rose. 2006. "Biopower today." *BioSocieties* 1:195–217.

"Race of monsters seen: Pastor says hydrogen bomb is 'morally hideous.'" 1952. *New York Times*, December 8.

Raeburn, Paul. 1993. "Breast cancer risk seen as higher for lesbians." *Chicago Sun-Times*, February 5.

Raffensperger, Carolyn, and Joel Tickner. 1999. "Introduction: To foresee and forestall." In *Protecting Public Health and the Environment: Implementing the Precautionary Principle*, edited by Carolyn Raffensperger and Joel Tickner, 1–12. Washington: Island.

Rafter, Nicole Hahn. 1988. *White Trash: The Eugenic Family Studies, 1877–1919*. Boston: Northeastern University Press.

———. 2008. *The Criminal Brain: Understanding Biological Theories of Crime*. New York: New York University Press.

Rampton, Sheldon, and John Stauber. 2001. "Alar-Mists: How industry propaganda clouds the history of a 'health hoax.'" *Extra!* 14 (January–February): 26–28.

Randall, Thomas C., and Katrina Armstrong. 2003. "Differences in treatment and outcome between African-American and white women with endometrial cancer." *Journal of Clinical Oncology* 21 (22): 4200–06.

Rapp, Rayna. 1998. "Refusing prenatal diagnosis: The meanings of bioscience in a multicultural world." *Science, Technology, and Human Values* 23 (1): 45–70.

Reardon, Jenny. 2004. "Decoding race and human difference in a genomic age." *Differences* 15 (3): 38–65.

Rebbeck, Timothy R., et al. 2002. "Prophylactic oophorectomy in carriers of BRCA1 or BRCA2 mutations." *New England Journal of Medicine* 346 (21): 1616–21.

Reilly, Phillip. 1987. "Involuntary sterilization in the United States: A surgical solution." *Quarterly Review of Biology* 62 (2): 153–70.

Risch, Harvey A., et al. 2001. "Prevalence and penetrance of germline BRCA1 and BRCA2 mutations in a population series of 649 women with ovarian cancer." *American Journal of Human Genetics* 68 (3): 700–710.

Risch, Neil, Esteban Burchard, Elad Ziv, and Hua Tang. 2002. "Categorization of humans in biomedical research: Genes, race, and disease." *Genome Biology* 3 (7): 1–12.

Roberts, Dorothy. 1997a. *Killing the Black Body: Race, Reproduction, and the Meaning of Liberty*. New York: Pantheon.

———. 1997b. "The nature of Blacks' skepticism about genetic testing." *Seton Hall Law Review* 27: 971–79.

Roberts, J. Timmons, and Melissa M. Toffolon-Weiss. 2001. *Chronicles from the Environmental Justice Frontline*. Cambridge: Cambridge University Press.

Roberts, Stephanie A., et al. 1998. "Differences in risk factors for breast cancer: Lesbian and heterosexual women." *Journal of the Gay and Lesbian Medical Association* 2 (3): 93–101.

———. 1999. "Heritability of cellular radiosensitivity: a marker of low-penetrance predisposition in breast cancer?" *American Journal of Human Genetics* 65 (3): 784–94.

Rocca, Walter A., et al. 2006. "Survival patterns after oophorectomy in premenopausal women: A population-based cohort study." *Lancet Oncology* 7 (10): 821–28.

Rochelle, Anne. 1994. "Breast cancer discovery elates scientists." *Atlanta Journal-Constitution*, September 15.

Rockhill, Beverly. 2001. "The privatization of risk." *American Journal of Public Health* 91 (3): 365–68.

Romualdi, Chiara, et al. 2002. "Patterns of human diversity, within and among continents, inferred from biallelic DNA polymorphisms." *Genome Research* 12 (4): 602–12.

Rose, Nikolas. 2006. *The Politics of Life Itself*. Princeton: Princeton University Press.

Rosen, Barry, et al. 2004. "Systematic review of management options for women with a hereditary predisposition to ovarian cancer." *Gynecologic Oncology* 93 (2): 280–86.

Rosner, David, and Gerald Markowitz. 2005. "Standing up to the lead industry:

An interview with Herbert Needleman." *Public Health Reports* 120 (May–June): 330–37.

Rothenberg, Karen. 1997. "Miracles of genetics can bear heavy cost: Participants in tests sometimes lose privacy and health insurance." *Baltimore Sun*, July 20.

Rouse, Joseph. 2004. "Feminism and the social construction of scientific knowledge." In *The Feminist Standpoint Theory Reader: Intellectual and Political Controversies*, edited by Sandra Harding, 353–74. New York: Routledge.

Rubin, Stephen C. 2003. "BRCA-related ovarian carcinoma: Another piece of the puzzle?" *Cancer* 97 (9): 2127–29.

Runowicz, Carolyn D. 1999. "Genetic susceptibility to breast and ovarian cancer: Assessment, counseling and testing guidelines. Appendix VIII: Prophylactic oophorectomy." New York State Department of Health. http://www.health.ny.gov/diseases/cancer/obcancer/contents.htm (accessed October 30, 2012).

Russo, J., L. K. Tay, and I. H. Russo. 1982. "Differentiation of the mammary gland and susceptibility to carcinogenesis." *Breast Cancer Research and Treatment* 2 (1): 5–73.

Rutkow, Ira M. 1999. "Ephraim McDowell and the world's first successful ovariotomy." *Archives of Surgery* 134 (8): 902.

Saltus, Richard. 1994. "Mutated gene tied to early breast cancer is located." *Boston Globe*, September 15.

Samuels, Suzanne. 1996. "The fetal protection debate revisited: The impact of U.A.W. v. Johnson Controls on the federal and state courts." *Women's Rights Law Reporter* 17:209.

Sandler, Ronald, and Phaedra C. Pezzullo, eds. 2007. *Environmental Justice and Environmentalism: The Social Justice Challenge to the Environmental Movement.* Cambridge: MIT Press.

Sanz, David J., et al. 2010. "A high proportion of DNA variants of BRCA1 and BRCA2 is associated with aberrant splicing in breast/ovarian cancer patients." *Clinical Cancer Research* 16 (6): 1957–67.

Schemo, Diana Jean. 1994. "L.I. breast cancer is possibly linked to chemical sites." *New York Times*, April 13.

Schrag, Deborah, Karen M. Kuntz, Judy E. Garner, and Jane C. Weeks. 1997. "Decision analysis—Effects of prophylactic mastectomy and oophorectomy on life expectancy among women with BRCA1 and BRCA2 mutations." *New England Journal of Medicine* 336 (20): 1465–71.

Schwartz, David, and Francis Collins. 2007. "Environmental biology and human disease." *Science* 316 (May 4): 695–96.

"The Secret of Life." 1958. *Time*, July 14, 50–54.

Sellers-Diamond, Alfreda A. 1994. "Disposable children in black faces: The Violence Initiative as inner-city containment policy." *UMKC Law Review*, 62 (Spring): 423.

Seltzer, Vicki, et al. 1995. "Ovarian cancer: Screening, treatment, and follow-up.

NIH Consensus Development Panel on Ovarian Cancer." *Journal of the American Medical Association* 273 (6): 491–97.

Selvin, Barbara. 1993. "One in three lesbians risks death from breast cancer." *London Guardian*, February 6.

Sengoopta, Chandak. 2000. "The modern ovary: Constructions, meanings, uses." *History of Science* 38:425–88.

Shapiro, Robert. 1991. *The Human Blueprint: The Race to Unlock the Secrets of Our Genetic Script*. New York: St. Martin's.

Shapiro, S. 2006. "Does retention of the ovaries improve long-term survival after hysterectomy? The validity of the epidemiological evidence." *Climacteric* 9 (3): 161–63.

Sharp, Richard R., and Carl Barrett. 2000. "The environmental genome project: Ethical, legal, and social implications." *Environmental Health Perspectives* 108 (4): 279–81.

Sherwin, Susan. 1992. *No Longer Patient: Feminist Ethics and Healthcare*. Philadelphia: Temple University Press.

Shildrick, Margrit. 2004. "Genetics, normativity, and ethics: Some bioethical concerns." *Feminist Theory* 5 (2): 149–65.

Shim, Janet K. 2002. "Understanding the routinised inclusion of race, socioeconomic status and sex in epidemiology: The utility of concepts from technoscience studies." *Sociology of Health & Illness* 24 (2): 129–50.

Shostak, Sara. 2010. "Marking populations and persons at risk: Molecular epidemiology and environmental health." In *Biomedicalization: Technoscience, Health, and Illness in the U.S.*, edited by Adele E. Clarke et al., 242–62. Durham: Duke University Press.

Siegel, Judy. 1996. "Ashkenazi cancer gene also found in Iraqis." *Jerusalem Post*, May 2.

Slovut, Gordon. 1995. "Gene linked to breast, ovarian cancer." *Minneapolis Star Tribune*, October 26.

Smith, Tasha R., et al. 2003. "DNA-repair genetic polymorphisms and breast cancer risk." *Cancer, Epidemiology, Biomarkers, and Prevention* 12:1200–4.

———. 2008. "Polygenic model of DNA repair genetic polymorphisms in human breast cancer risk." *Carcinogenesis* 29 (11): 2132–38.

Smith-Rosenberg, Carroll, and Charles E. Rosenberg. 1999. "The female animal: Medical and biological views of woman and her role in nineteenth century America." In *Women and Health in America: Historical Readings*, edited by Judith Walzer Leavitt, 111–30. 2nd ed. Madison: University of Wisconsin Press.

Solinger, Rickie. 1992. *Wake Up Little Susie: Single Pregnancy and Race before Roe V. Wade*. New York: Routledge.

———. 2001. *Beggars and Choosers: How the Politics of Choice Shapes Adoption, Abortion, and Welfare in the United States*. New York: Hill and Wang.

Sorelle, Ruth. 1995. "Bigger role suspected for breast cancer gene." *Houston Chronicle*, November 3.

Spear, Scott, et al. 1999. "Prophylactic mastectomy, oophorectomy, hysterectomy, and immediate transverse rectus abdominis muscle flap breast reconstruction in a BRCA-2 positive patient." *Plastic and Reconstructive Surgery* 103 (2): 548–53.

Stabile, Carol A. 1992. "Shooting the mother: Fetal photography and the politics of disappearance." *Camera Obscura* 10 (128): 178–205.

———1995. "Resistance, recuperation, and reflexivity: The limits of a paradigm." *Critical Studies in Mass Communication* 12 (4): 403–33.

Steeg, Patricia S. 1996. "Granin expectations in breast cancer?" *Nature Genetics* 12 (March): 223–25.

Stefanic, Jean, and Richard Delgado. 2000. *Critical Race Theory: The Cutting Edge.* Philadelphia: Temple University Press.

Steingraber, Sandra. 1998. *Living Downstream: A Scientist's Personal Investigation of Cancer and the Environment.* Reading, MA: Addison-Wesley.

———. 2001. *Having Faith: An Ecologist's Journey to Motherhood.* New York: Berkley.

Stellman, Steven D., et al. 2000. "Breast cancer risk in relation to adipose concentrations of organochlorine pesticides and polychlorinated biphenyls in Long Island, New York." *Cancer Epidemiology, Biomarkers, and Prevention* 9 (November): 1241–49.

Sternberg, Stephen S. 1994. "Re: DDT and breast cancer." *Journal of the National Cancer Institute* 86 (14): 1094.

Stevens, Jacqueline. 2002. "DNA and other linguistic stuff." *Social Text* 20 (1): 105–36.

———. 2003. "Racial meaning and scientific methods: Changing policies for NIH-sponsored publications reporting human variation." *Journal of Health Politics, Policy and Law* 28 (6):1033–87.

———. 2008. "The feasibility of government oversight of NIH-funded population genetics." In *Revisiting Race in a Genomic Age*, edited by Barbara A. Koenig, Sandra Soo-Jin Lee, and Sarah S. Richardson, 320–41. New Brunswick: Rutgers University Press.

Struewing, Jeffery P., et al. 1995a. "The carrier frequency of the *BRCA1* mutation is approximately 1 percent in Ashkenazi Jewish individuals." *Nature Genetics* 11 (October): 198–200.

———. 1995b. "Detection of eight *BRCA1* mutations in 10 breast/ovarian cancer families, including one family with male breast cancer." *American Journal of Human Genetics* 57:1–7.

———. 1995c. "Prophylactic oophorectomy in inherited breast/ovarian cancer families." *Journal of the National Cancer Institute Monographs* (17): 33–35.

———. 1997. "The risk of cancer associated with specific mutations of BRCA1 and BRCA2 among Ashkenazi Jews." *New England Journal of Medicine* 336 (20): 1401–8.

Studd, John. 1989. "Prophylactic oophorectomy." *British Journal of Obstetrics and Gynaecology* 96 (5): 506–9.

———. 2006a. "Does retention of the ovaries improve long-term survival after hysterectomy? Prophylactic oophorectomy." *Climacteric* 9 (3): 164–66.

———. 2006b. "Ovariotomy for menstrual madness and premenstrual syndrome—19th-century history and lessons for current practice." *Gynecological Endocrinology* 22 (8): 411–15.

Sunder Rajan, Kaushik. 2006. *Biocapital: The Constitution of Postgenomic Life.* Durham: Duke University Press.

Sutton, Larry. 1996. "Gov inks cancer study: New law will probe link to pesticides." *New York Daily News,* July 9.

Swirsky, Joan. 1990. "The latest study on breast cancer leaves L.I. women still in the dark." *New York Times,* July 15.

Tang, Hua, et al. 2005. "Genetic structure, self-identified race/ethnicity, and confounding in case-control association studies." *American Journal of Human Genetics* 76 (2): 268–75.

Tesh, Sylvia Noble. 2000. *Uncertain Hazards: Environmental Activists and Scientific Proof.* Ithaca: Cornell University Press.

Thiery, Michel. 1998. "Battey's operation: An exercise in surgical frustration." *European Journal of Obstetrics and Gynecology and Reproductive Biology* 81 (2): 243–46.

Thompson, Hayley S., et al. 2002. "Psychosocial predictors of BRCA counseling and testing decisions among urban African-American women." *Cancer Epidemiology, Biomarkers, and Prevention* 11 (12): 1579–85.

Thompson, Marilyn, et al. 1995. "Decreased expression of BRCA1 accelerates growth and is often present during sporadic breast cancer progression." *Nature Genetics* 9 (April): 444–50.

Tobacman, Joanne K., et al. 1982. "Intra-abdominal carcinomatosis after prophylactic oophorectomy in ovarian-cancer-prone families." *Lancet* 2 (8302):795–97.

Toh, Sengwee, et al. 2010. "Coronary heart disease in recipients of postmenopausal estrogen plus progestin therapy: Does the increased risk ever disappear? A randomized trial." *Annals of Internal Medicine* 152 (4): 211–17; W47–W52.

Toniolo, Paolo, and Ikuko Kato. 1996. "Jewish religion and risk of breast cancer." *Lancet* 348 (9029): 760.

"Tracking down chemical suspects." 2002. *Silent Spring Review,* Winter. http://www.silentspring.org/pdf/our_publications/SSIWinterReview02.pdf (accessed November 9, 2012). On file with author.

Treichler, Paula. 1999. *How to Have Theory in an Epidemic: Cultural Chronicles of AIDS.* Durham: Duke University Press.

———, Lisa Cartwright, and Constance Penley, eds. 1998. *The Visible Woman: Imaging Technologies, Gender, and Science.* New York: NYU Press.

Trock, Bruce J. 1996. "Breast cancer in African American women: Epidemiology and tumor biology." *Breast Cancer Research and Treatment* 40 (1): 11–24.

Trumbull, Robert. 1955. "Atom survivors usually normal." *New York Times,* May 31.

Tusiani, Bea. 1998. "Growth pains on breast cancer." *New York Times*, March 15.

Tuttle, Todd M., et al. 2009. "Increasing rates of contralateral prophylactic mastectomy among patients with ductil carcinoma in situ." *Journal of Clinical Oncology* 27 (9): 1362–67.

U.S. Department of Health and Human Services. 2006. "Two NIH Initiatives Launch Intensive Efforts to Determine Genetic and Environmental Roots of Common Diseases." http://www.nih.gov/news/pr/feb2006/nhgri-08.htm (Accessed November 26, 2012).

Utian, Wulf H. 1999. "Historical perspectives in menopause: An historical perspective of natural and surgical menopause." *Menopause* 6 (2): 83–86.

Van Dijck, José. 1998. *Imagenation: Popular Images of Genetics*. New York: New York University Press.

Van Nagell, J. R., Jr. 1991. Editorial. *Gynecologic Oncology* 43 (2): 89–91.

Waddell, Craig. 1990. "The role of pathos in decision-making process: A study in the rhetoric of science policy." *Quarterly Journal of Speech* 76 (4): 381–400.

Wade, Nicholas. 2003. "Once again, scientists say human genome is complete." *New York Times*, April 15.

Wagner, Wendy. 1995. "The science charade in toxic risk regulation." *Columbia Law Review* 95 (7): 1613–1723.

Waldby, Catherine. 1996. *AIDS and the Body Politic: Biomedicine and Sexual Difference*. London: Routledge.

——— hapand Melinda Cooper. 2008. "The biopolitics of reproduction: Post-Fordist biotechnology and women's clinical labour." *Australian Feminist Studies* 23 (55): 57–73.

Waldby, Catherine, and Robert Mitchell. 2006. *Tissue Economies: Blood, Organs, and Cell Lines in Late Capitalism*. 2nd ed. Durham: Duke University Press.

Wallach, Robert C. 2005. "Ovarian conservation at the time of hysterectomy for benign disease." *Obstetrics and Gynecology* 106 (5): 1106–7.

Wassermann, M., et al. 1976. "Organochlorine compounds in neoplastic and adjacent apparently normal breast tissue." *Bulletin of Environmental Contamination and Toxicology* 15 (4): 478–84.

Watson, James. 1968. *The Double Helix: A Personal Account of the Discovery of the Structure of DNA*. New York: Atheneum.

——— and Francis Crick. 1953a. "Genetical implications of the structure of deoxyribonucleic acid." *Nature* 171 (May 30): 964–67.

———. 1953b. "Molecular structure of nucleic acids: A structure for deoxyribose nucleic acid." *Nature* 171 (April 25): 737–38.

Weinberg, Robert A. 1987. "The case against gene sequencing." *Scientist*, November 16, 11.

Weis, Brenda K., et al. 2005. "Personalized exposure assessment: Promising approaches for human environmental health research." *Environmental Health Perspectives* 113 (7): 840–48.

Weitzel Jeffrey N., et al. 2007. "Evidence for common ancestral origin of a recurring BRCA1 genomic rearrangement identified in high-risk Hispanic families." *Cancer Epidemiology, Biomarkers, and Prevention* 16 (8): 1615–20.

West, Emelda, and Pat Melancon. 2002. Untitled presentation at the annual meeting of the National Communication Association, New Orleans, November.

White, John H. 1998. "Breast cancer risk higher in lesbians." *Chicago Sun-Times* October 16.

Wiggam, Albert E. 1922. "The new decalogue of science: An open letter from the biologist to the statesman." *Century* 103:643–50.

———. 1926. "The rising tide of degeneracy: What everybody ought to know about eugenics." *World's Work* 53 (1): 25–33.

Wiggam, Albert E. 1922. "The new decalogue of science: An open letter from the biologist to the statesman." *Century Magazine* 103: 643–50.

Wiley, Harvey. 1919. "Making the New American." *Good Housekeeping*, September, 68, 164–67.

Willett, Walter C. 2002. "Balancing life-style and genomics research for disease prevention." *Science* 296 (5568): 695–98.

Wills, Christopher. 1991. *Exons, Introns, and Talking Genes: The Science behind the Human Genome Project*. New York: Basic.

Wilson, James F., et al. 2001. "Population genetic structure of variable drug response." *Nature Genetics* 29 (November): 265–69.

Wingerson, Lois. 1991. *Mapping Our Genes: The Genome Project and the Future of Medicine*. New York: Plume.

Wolff, Mary S., Julie A. Britton, and Valerie P. Wilson. 2003. "Environmental risk factors for breast cancer among African-American women." *Cancer* 97 (1 Supplement): 289–310.

Wolff, Mary S., et al. 1993. "Blood levels of organochlorine residues and risk of breast cancer." *Journal of the National Cancer Institute* 85 (9): 648–52.

Wooster, Richard, et al. 1995. "Identification of the breast cancer susceptibility gene BRCA2." *Nature* 378 (December 21): 789–92.

Wright, Robert. 1990. "Achilles' helix." *New Republic*, July, 9–16.

———. 1995. "The biology of violence." *New Yorker*, March 13, 68-77.

Wright, Susan. 1994. *Molecular Politics: Developing American and British Regulatory Policy for Genetic Engineering, 1972–1982*. Chicago: University of Chicago Press.

Wynne, Brian. 1993. "Public uptake of science: A case for institutional reflexivity." *Public Understanding of Science* 2 (4): 321–37.

Yalom, Marilyn. 1998. *History of the Breast*. New York: Ballantine.

Abortion, and prenatal genetic testing, 217n184

ACOG. *See* American Congress of Obstetricians and Gynecologists

Activism: epidemiology and, 174, 241n176; feminist health activists, 82, 188. *See also* Breast cancer activists; Environmental justice activism

Africans: ancestry of African Americans and, 123; genetic diversity, 126–27; migration, 224n85

African Americans, 185; ancestry of Africans and, 123; chemicals and, 148; ecosocial model and, 131–32; genetic diversity of, 126–27; pollution and, 162–64, 236n126; Reagan administration and, 47; sickle cell and, 117–18, 148, 186, 230n34. *See also* Blacks; Race

African American women, 17–18, 103, 107, 192n54; African women, breast cancer and, 115–19, 222n57; ancestry, 127; BRCA mutations in, 125–26, 242n10; breast cancer and, 114–21, 125–26, 222n57; breast cancer research on, 109, 116–17, 220n23. *See also* Black women

African women: African American women, breast cancer and, 115–19, 222n57; ancestry, 115; founder mutations in, 115

AIDS, 9, 189n3

Alar, 146, 229n26, 238n153

American Congress of Obstetricians and Gynecologists (ACOG), 76, 77, 211n109

American Eugenics Society, 25, 27, 30, 193n7

American Journal of Human Genetics, 35

American Society of Human Genetics, 35

Ancestry: African American and African, 123; African American women, 127;

African women, 115; Ashkenazi, 120; in founder mutations, 115; Hispanic, 120–21; population and, 115; race and, 115, 117, 136; shared, 117; of white women, 119–20, 128, 223n74

"Annual Report to the Nation on Cancer," 128

Ashkenazi ancestry, 120

Ashkenazi Jews, and BRCA mutations, 104, 105, 109, 113–14, 221nn43–44, 240n165

Ataxia-telangiectasia (AT), 156, 234nn81–82, 234nn85–86

ATM gene, 156, 234n86

AT mutated (ATM), 156

Balaban, Barbara, 172, 173, 204n195, 240n167

Baltimore, David, 45, 198n121

Barish, Geri, 204n197

Battey, Robert, 207n44

Battey's operation, 71, 207n44

Beck, Ulrich, 167–68

Benzene, 147, 168, 229n30, 230n32

Biocapital, 49; capitalism and, 48; genes and, 52; social, economic order and, 48

Bioethics, 183–84, 242n14; breast cancer, 57–59, 180; conventional, 181; feminist, 181, 182, 183, 186; genetic testing and, 59, 181, 205n207; genomic medicine and, 57; heredity and, 182; limitations of/limits to, 182–86; race and, 185–86; research, 57–58, 204n202

Biologism, 141–42

Biology: capitalism and, 8–9; DNA and, 49–50; genes and, 49–50; genetics and, 118, 223n65; genomics and, 48, 200n140. *See also* Molecular biology

Biomedicine, 10–11, 14, 48

ABOUT THE AUTHOR

Kelly E. Happe is an Assistant Professor in the Department of Communication Studies and the Institute for Women's Studies at the University of Georgia.